Understanding
Electro-Mechanical
Engineering

JUL – 8 1999

IEEE PRESS Understanding Science & Technology Series

The IEEE PRESS Understanding Science & Technology Series treats important topics in science and technology in a simple and easy-to-understand manner. Designed expressly for the nonspecialist engineer, scientist, or technician as well as the technologically curious—each volume stresses practical information over mathematical theorems and complicated derivations.

Books in the Series

Deutsch, S., *Understanding the Nervous System: An Engineering Perspective*

Evans, B., *Understanding Digital TV: The Route to HDTV*

Hecht, J., Sr., *Understanding Lasers: An Entry-Level Guide,* Second Edition

Kamm, L., *Understanding Electro-Mechanical Engineering: An Introduction to Mechatronics*

Kartalopoulos, S. V., *Understanding Neural Networks and Fuzzy Logic: Basic Concepts and Applications*

Nellist, J. G., *Understanding Telecommunications and Lightwave Systems: An Entry-Level Guide,* Second Edition

Sigfried, S., *Understanding Object-Oriented Software Engineering*

Ideas for future topics and authorship inquiries are welcome. Please write to IEEE PRESS: The Understanding Series

Understanding Electro-Mechanical Engineering

An Introduction to Mechatronics

Lawrence J. Kamm

formerly President of MOBOT Corp.

The Institute of Electrical and Electronics Engineers, Inc., New York

IEEE PRESS
445 Hoes Lane, P.O. Box 1331
Piscataway, NJ 08855-1331

Editorial Board
J. B. Anderson, *Editor in Chief*

S. Blanchard	R. F. Hoyt	R. S. Muller
R. S. Blicq	S. V. Kartalopoulos	I. Peden
M. Eden	P. Laplante	W. D. Reeve
R. Herrick	J. M. F. Moura	E. Sánchez-Sinencio
G. F. Hoffnagle		

Dudley R. Kay, *Director of Book Publishing*
Carrie Briggs, *Administrative Assistant*
Lisa S. Mizrahi, *Review and Publicity Coordinator*

Technical Reviewers

Ali Seireg
Ebaugh Chair Professor, University of Florida, Gainesville
Kaiser Chair Professor, University of Wisconsin, Madison

Dennis Lieu
Professor of Mechanical Engineering, University of California, Berkeley

Tzyh-Jong Tarn
Professor, Systems Science and Mathematics, Washington University

Charles Mischke, Ph.D.
Emeritus Professor of Mechanical Engineering, Iowa State University

Seymour Kremen, Ph.D.
Consultant

This book may be purchased at a discount from the publisher when
ordered in bulk quantities. For more information, contact:

IEEE PRESS Marketing
Attn: Special Sales
P.O. Box 1331
445 Hoes Lane
Piscataway, NJ 08855-1331
Fax: (908) 981-9334

© 1996 by the Institute of Electrical and Electronics Engineers, Inc.
345 East 47th Street, New York, NY 10017-2394

All rights reserved. No part of this book may be reproduced in any form,
nor may it be stored in a retrieval system or transmitted in any form,
without written permission from the publisher.

Printed in the United States of America

10 9 8 7 6 5 4 3 2 1

ISBN 0-7803-1031-4
IEEE Order Number: PP3806

Library of Congress Cataloging-in-Publication Data

Kamm, Lawrence J.
    Understanding electro-mechanical engineering : an introduction to
mechatronics / by Lawrence J. Kamm
        p. cm.
    Includes bibliographical references and index.
    ISBN 0-7803-1031-4
    1. Mechatronics. 2. Electromechanical devices. I. Title.
TJ163.12.K35   1995
621—dc20   95-16524
                                                                CIP

To Edith, Dorie, Ron, Megan, and Jeffrey

Brief Contents

Contents

Chapter 2   Mechanical Engineering   35

PART II   Understanding the Devices and Systems   65

Chapter 3   Generators and Motors   67

Chapter 5   Amplifiers   101

Chapter 6   Actuators   111

List of Illustrations

Preface

A rocket streaks across the sky, guided by electro-mechanical gyroscopes and accelerometers. A cardiologist diagnoses a patient, helped by an electro-mechanical electro-cardiograph. A scientist explores a law of nature on a computer, its data stored on electro-mechanical tape and disk memories. And the cardiologist and the scientist eat breakfasts prepared with electro-mechanical toasters.

In the real world, mechanical engineering and electrical engineering are inextricably entwined. Every electrical device is a mechanical device designed for its electrical properties and manufactured in a factory of mechanical machines; many mechanical devices are partly electrical and most are made by machines that are electrically powered and electrically controlled. Electro-mechanical engineering is interdisciplinary. Electro-mechanical engineering deals with devices and systems combining electrical and mechanical phenomena. The art has been growing for over a century, but only recently has a new name—mechatronics—been applied, whence the title of this book [A54].

A voltage amplifier is purely electrical; a steam engine is purely mechanical; but a motor, solenoid, switch, computer disk drive, and autopilot are electro-mechanical.

This book is for engineers who are already involved with electro-mechanical devices and systems and for those who only want to broaden their general knowledge. It is for both electrical and mechanical engineers and for engineering students who are considering entering the field. Because the book contains almost no mathematics, it can be understood by technicians and laypersons who do not have formal engineering training. It will help engineering managers to understand and judge what their people are up to. It will help you to use electro-mechanical devices in your work, and it may even tempt you into venturing into the field as a profession.

The essays on design apply to engineering design in general, not just to the design of electro-mechanical devices and systems.

This book is a qualitative survey of principles and examples. It contains almost no performance or size data except some suggestions of magnitude; there are many orders of magnitude in the electrical and mechanical parameters involved and in the accuracy of the devices dealt with.

Many an electro-mechanical device is designed by a team of electrical and mechanical engineers, each understanding little of the capabilities and limitations of the other's art. The result can suggest the classic definition of a camel as a horse designed by a committee. But if both arts reside in the same mind, both arts can be combined in an optimum design.

I learned this as a young electrical engineer in the relay business. After endless hours of wrangling with a mechanical designer with whom I shared mutual ignorance, I sat myself down at a drawing board and the result was a revelation.

Scattered through the book are essays on systems, devices, history, design, and engineering philosophies that relate to the subject matter. If my statements are controversial, so much the better.

I hope my experience will stretch your imagination and help your career—and do so with interesting and easy reading.

Lawrence J. Kamm

About the Author

Lawrence J. Kamm's formal education includes a B.S.E.E. from Columbia University in 1941 and an M.E.E. from Brooklyn Polytechnic Institute in 1946. He has Professional Engineer licenses from New York, Maryland, and California and is a Registered Patent Agent and a Certified Manufacturing Engineer (SME). He is a Distinguished Lecturer at Milwaukee School of Engineering.

The author belongs to Sigma Xi, the Institute of Electrical and Electronics Engineers (IEEE) and, when he worked in those fields, the American Rocket Society (now the American Institute of Aeronautics and Astronautics), the Society of Manufacturing Engineers (SME), and the Robot Institute.

He holds 38 patents and has published or presented 26 papers and much trade literature. He is the author of the books *Successful Engineering* (republished by IEEE Press as *Real-World Engineering*) and *Designing Cost-Efficient Mechanisms*.

Kamm has invented and designed electro-mechanical devices of great diversity. These include robots, numerical controls, computer peripherals, space vehicles and components, simulators, a mail-sorter memory, a heart-lung machine, pacemaker components, automatic test-and-sort and other manufacturing equipment, transducers, switch gear, and engine components.

He has worked as an employed engineer in both small and large companies, as an entrepreneur (Numerical Control Corporation, Devonics, Inc., Typagraph Corporation, MOBOT Corporation), as a teacher of design theory and practice, and as a consulting engineer and inventor. Consulting and writing are his present activities. He lives in San Diego.

Other books by L. J. Kamm

Real-World Engineering
(Originally: *Successful Engineering*)

Designing Cost-Efficient Mechanisms

Introduction

Most academic curricula and professional society literature ignore electro-mechanical devices other than generators, motors, and servo theory. This may be because:

1. Their designs are diverse and may have strange-looking structures.
2. Their engineering is based largely on judgment, inventiveness, and experimentation as well as on mathematical analysis.
3. They fit partly on another society's turf (ASME, IEEE, SME, SAE, AIAA, or other.)

At the moment, there is no jurisdictional dispute over which electro-mechanical device belongs to which engineering society. This book's working definition of an electro-mechanical device or system is any that combines electrical and mechanical technologies.

Since those hyphenated seven syllables, *electro-mechanical,* would appear many times in this book and would slow your reading they are abbreviated to EM. EM device, EM system, EM component, and the like are abbreviated to EMD.

Some IEEE professional societies include certain EMDs used in their particular fields. Examples are antennas, electronic packaging, servomechanisms, and some power distribution devices, but for the most part it is their electrical characteristics that are studied. An Electro-Mechanical Engineering Society would stimulate the publication of papers on advances in EM engineering instead of leaving such publication to commercial catalogs and to patents. A great deal of non-confidential knowledge lies in companies' internal manuals. It is not available to those not employed in those companies because there is no society that recognizes this knowledge as belonging in its purview. If there were an Electro-Mechanical Engineering Society, many engineers would find a home.

We form warring factions in politics, sports, and professions, but to claim superiority of one branch of engineering over another is juvenile. The case in point is competition between electronics and mechanics.

The younger art of electronics has superseded the older art of mechanics in computing, in most information handling, in sound reproduction, and in controlling mechanical actions. It has made phenomenal achievements in areas where there is no mechanical competition at all. The enormous success of electronics has fed the professional pride of its practitioners.

Mechanics is a much older art—several thousand years older if we go to its beginnings. Therefore its achievements are taken for granted, and it has made no recent breakthroughs comparable to those of electronics.

The *combination* of the two arts, EM engineering, has produced tape and disk memories, servos, and guidance systems, to name a very few. Much electronic technology helps mechanisms to produce non-electronic effects such as music reproduction and mechanical positioning. Conversely, much mechanical technology, usually electronically controlled, manufactures electronic components such as integrated circuits, and much is part of electronic systems such as computers and large pointable antennas. The combination, EM engineering, is what this book is about.

                           *   *   *

Electrical and mechanical things are both simple and complicated in different ways. All electrical circuits (but not fields) can be fully represented by diagrams in a single plane. Even most physical embodiments lie in a small number of planes with fewer connections from plane to plane than within each plane. These embodiments include printed circuits, panel boards, and the integrated circuits that are the most complex artifacts ever made by humankind.

The mechanical simplicity of electrical circuits, as wired interconnections among mechanically simple components, has made it economical to design and build systems of enormous complexity and capability. Some make an internal combustion engine trivially simple in comparison. (But not simpler in quantitative analysis and design!)

On the other hand most mechanical devices are three-dimensional and need three-dimensional parts and features to make them work. (The art of mechanical drafting is the art of representing three-dimensional subjects on two-dimensional paper; the same is true of the aesthetic arts of representational drawing and painting.)

Almost nothing can be learned about an electrical circuit just by looking at the physical hardware; it is mysterious unless studied. On the other hand most mechanical devices are superficially obvious; it is their subtleties that require study. This superficial "obviousness" leads some to condescend to mechanical engineering as an inferior art.

One usually thinks of EMDs as having moving parts, and most do. However every static electrical component is a mechanical structure designed for its electrical properties. A field of EM engineering deals with the design and manufacture of electrical components and assemblies. Solid-state electronics is the most dramatic example, but the statement is equally true of tiny resistors, big radio transmitters, and giant transmission lines.

*   *   *

Part 1 of *Understanding Electromechanical Engineering* is a review of the electrical and mechanical technologies on which EMDs are based so that you will better understand the EMDs themselves.

Part 2 is a compendium of EMDs, small, big, simple, complex, rough, and incredibly fine. You will be surprised at the number of kinds there are. Examples were selected to illustrate principles and to describe components that are combined into working hardware. The intent was to enlarge your understanding rather than to give you an exhaustive catalog of devices. Such a catalog would include all the devices in regular use, all the devices which have been used but are now obsolete, and all the device ideas described in the patent literature but were never exploited. The examples given show only principles and the breadth of the field; a complete survey of all EMDs would be an encyclopedia.

Part 3 is a series of essays on designing EMDs, on the design process in general, and on your design career.

*   *   *

If you select a ready-made device for use, you will use published specifications, perhaps perform mathematical analysis of the systems for which you chose the device, and you may confirm the published data with tests. But if you *design* an EMD, you will perform a great deal of calculation, some simple and some quite complex, in addition to your qualitative thinking and judgment. The References and Bibliography at the end of this book will give you a start on such analyses.

All the mathematics needed for EM engineering has long been treated in textbooks of physics, electrical engineering, and mechanical engineering; it would be wasteful to repeat it here. Some of these texts are listed in the References and Bibliography at the end of the book. Specific mathematical analysis of motors, generators, and magnetic circuits is also available in many textbooks. Computer software is now available for accurate treatment of the shape and parameters of magnetic, thermal, and stress fields, which used to be dealt with only by rough approximations. The only mathematics in this book deal with the basics of EM force.

Not included in the book are thousands of kinds of primarily mechanical machinery that are electrically powered and controlled, although they are EMDs by definition. Nor are there included primarily electronic devices which may contain a loudspeaker or a power switch and which, therefore, are also EMDs by definition. This book is confined to those EMDs which are components of those systems (motors, switches, clutches, loudspeakers, and so forth) and to systems in which electrical and mechanical technologies are closely interrelated, such as servo-mechanisms.

There exist many variations of each type of device described. This book is limited to typical examples; for a comprehensive study of variations, see manufacturers' catalogs [Reference E1] and the patents in the field.

Most of the source literature in the field is catalogs and patent files. Fortunately most catalogs are carefully written to be easily understood in order to encourage you to buy. Some catalogs are so well written that they are valuable textbooks, and some are even labeled as "Handbook of ——."

                            *   *   *

There is an enormous industry making EMDs. If you want to study a class of device more deeply than this survey provides, you should start with a directory such as *Thomas Register* [Reference E1]. Then collect catalogs and consult sales engineers. The author has started most of his design projects—including finding his first job—in *Thomas.*

Manufacturing companies range in size from Fortune 100 corporations to tiny specialty shops. Most of the latter were started by inventive entrepreneurs who spun off from big companies, government, or academia. There is no correlation between a manufacturer's size and its products' merit.

Many standards and specifications are published, but they assume a preexisting knowledge of the products involved. However they do comprise handbooks for designers. Some major sources are listed in Group G of the References and Bibliography at the end of this book.

                            *   *   *

Regarding the design of the book:

It was originally intended to provide photographs of the devices described, to get these photographs from manufacturers' catalogs, and to have the manufacturers' eager cooperation in exchange for the publicity. A richly illustrated book with free photography was anticipated. The cooperation was there, but the study of catalogs to choose photographs showed that most EMD photographs are utterly obscure in showing principles of operation.

Device geometry is largely a dense, three-dimensional packaging of device elements; looking at a picture of an assembled EMD is about as informative as looking at an assembled electrical circuit. The 3D visualization problem is so severe that even with the study of cutaway models of some machines in technology museums, it is still difficult to understand their structure and operation. Therefore the only illustrations are drawings which are intended to make up in clarity what they lack in detail.

The book avoids most numerical data such as size, range, accuracy, sensitivity, and so forth. because such data would probably antagonize some engineer or manufacturer by neglecting his particular achievements. You must again be referred to the catalog literature.

*   *   *

Some words about classification, *taxonomy*. Many of the devices described belong under two or more chapter headings. For example, a gyroscope is both a Transducer (Chapter 7) and a component of Military Devices (Chapter 15). Each category was chosen to associate the device with similar devices, and cross-references are given, but if you disagree with some of the category assignments it will not be at all surprising.

It seemed plausible that the U.S. Patent Office might already have devised a classification scheme that this book could use, but patent classification proved not useful for the book. The Appendix lists the Patent Office classes containing EMDs, each class having many subclasses, and sometimes subsubclasses, and each subclass having many patents, each with at least one idea legally new at the time of its filing.

The Constitution authorized the patent system in order to stimulate and to teach the applied arts; inventors are given a limited monopoly in exchange for teaching their inventions to the public. (The story goes that when von Braun, the chief engineer of the German V2 missile, was asked how he had advanced so much faster than we he answered, "I read the patents of your great Dr. Goddard, of course.")

Unfortunately, although the patent files were intended to be a textbook of technology, they make difficult reading. Their contents are organized and written to suit the processes of patent issue and infringement litigation rather than for teaching. As a result, most people who actually read patents are either inventors and patent attorneys, to differentiate their claims from the prior art, or designers, to avoid infringement, not those using them as a textbook.

One of the problems is inventors. Those fellows just won't stick to an existing scheme of things and they feel free to shoot off in new and unexpected directions without worrying about definitions or categories.

*   *   *

As material for the book was organized and reorganized new relationships appeared among devices. Each of you is already familiar with at least some of the devices and ideas in this book, but if the book gives you additional insight into such relationships, it will be a success for that reason alone.

Related to the subject of classification is the subject of definitions. From the standpoint of a dictionary writer, there should be rigorously exclusive and inclusive definitions of all technical terms so that they have exact meaning in communications. But engineers getting things done usually know perfectly well from context what a technical term means without an explicit definition. This is not an excuse for careless semantics, and certainly not for sloppy syntax, but it is an argument that engineers should not spend much time arguing about definitions. (An exception is the writing of laws, codes, specifications, and contracts where an ambiguous word can cause an expensive conflict. For these documents, a list of definitions of the terms as used in the document is essential when there is any possible ambiguity.) In this text, words that may be new to you and whose meaning is implied by the adjacent text are printed in italics.

The subject of the book is EMDs and their engineering, but it would be a disservice to you to ignore the existence of closely related devices that are non-electrical. Therefore there is mention of many of these related devices. For example, there are descriptions of electrically operated clutches and there is also mention of the host of non-electrically operated clutches.

Many devices have been made obsolete by later developments. For example, punched-tape memory was made obsolete by magnetic and semiconductor memories. However some concepts in these older devices may be useful in some future device that you may develop, they are of historical interest, and some are just plain fascinating to those who love the art, so the book describes many.

The book is broad at the expense of depth. The descriptions for each subject are intended to be sufficient to explain concepts without burdening you with details, but it was a matter of judgment how deep each should be. Each of you may want either more or less of any subject. Unquestionably some concepts and devices were overlooked; the only excuse is that the field is so large.

Now lets get into it.

PART I

Understanding the Science and Technology

Part 1 reviews the electrical engineering and mechanical engineering needed to understand electro-mechanical (EM) engineering at the level of this book.

Details and mathematical analyses are given in the References and Bibliography at the end of the book; references are identified by reference numbers in brackets [..]. Most of the subjects in Part 1 are extensively covered in the literature; the textbooks and articles cited in [Group A] being only a sampling of that literature. The handbooks, [Group B], are good places to start a literature search if you want to dig deeper.

The very first reference is always your own college textbook on the subject because you already understand it, or once did. Even high school textbooks provide adequate science background for much of this book.

Some examples of electro-mechanical devices (EMDs) are given in Part 1 to give a feeling of practical utility as well as of pure science and technology. More details about these examples, and many more examples, are given in the specialized chapters that follow in Part 2.

Since you have chosen to read this book, you probably already know at least some of the material in these two chapters. Nevertheless, please at least skim both chapters because you may find some ideas in them which are new to you or which will refresh your memory and help you understand the later chapters.

CHAPTER 1

Electrical Engineering

1.1 ELECTRICAL CONDUCTION

1.1.1 Solids

Electricity is conducted by some solids, including metals and carbon. Their resistivity is independent of current density or direction but is dependent on temperature.

Metal conductors are used in magnetizing coils, capacitor plates, resistors and heaters and as contacts, conductors, arc quenchers, mechanisms, and structures. Examples include contactors and circuit breakers [Chapter 9], [A9]. Magnetic metals are discussed below. In some relays, magnetic metals carry both magnetic flux and electric current.

Some solids are superconducting at cryogenic temperatures; that is, their resistivity is truly zero, so there is no heating from conducted current. Some recently developed ceramics are superconducting at higher temperatures and show promise for economical devices. This effect is extremely useful for electro-magnet coils carrying high current densities to generate intense magnetic fields. An example is the levitation systems in experimental high-speed trains, which, of course, are large EMDs.

Some solids are non-linear conductors in a variety of ways. The most important are doped semiconductors, such as silicon and germanium, used in transistors, diodes, light-emitting diodes (LEDs), photodiodes, and solid-state lasers. All these are used in EMDs. Silicon wafers are made into one-piece strain gauges and accelerometers. The conductors are built onto the surface and the entire disc serves as a diaphragm or proof mass, respectively. Silicon wafers are etched into parts for microminiature motors. Doped silicon is light sensitive and is the base of most photovoltaic cells.

Selenium is a semiconductor whose resistivity is sensitive to light and to current direction. In the *xerography* process, a uniform electro-static charge is deposited on the selenium surface of a rotating drum; an image is imposed, either by optical projection or by laser scanning; the charge is conducted away where light reaches the surface; powdered ink *(toner)* is applied to the surface and adheres to the remaining charged areas; the surface is pressed against paper and the ink is transferred; finally, heat melts the ink to form a permanent image [Chapter 11]. Xerography is used in photoprinters, computer laser printers, and some computer plotters. Selenium has also long been used in photoelectric cells and in rectifiers.

The resistivity of solids varies with temperature; in most metals, it increases almost linearly with absolute temperature. This is the basis for resistance thermometer temperature transducers, which measure temperature by measuring resistance [A18], [Chapter 7]. A few materials, such as carbon, have resistivities that decrease with increased temperature.

The volume of solids, liquids, and gases (at constant pressure) increases with temperature. This effect is used in a variety of devices including bimetal elements and other temperature-sensitive devices [Chapter 7].

Mechanical tensile and compressive stresses change the resistance of metal conductors. This effect has been of extreme value as the basis of the EM strain gauge [Chapter 7].

Carbon has unique properties of great value in EMDs. It does not melt, evaporate, or weld under ordinary conditions; thus it is valuable for electrodes and contacts. It is chemically almost inert except for oxidizing at high temperatures, and its oxides are gases that blow away, so its surface remains uncontaminated. (Once oxidized, carbon becomes the basic element of all organic compounds, of course.) [Chapters 3, 4, and 9].

Mercury is unique in being a metal that is liquid at room temperature; it melts at $-39°C$. Mercury has enabled the construction of contact makers that would otherwise not be possible [Chapter 9].

Properties of materials are tabulated in the handbooks listed in the References and Bibliography Group [B].

1.1.2 Liquids

Liquid conductors include molten metals such as mercury, aluminum, and iron. Mercury is particularly useful in switching devices in which liquid mercury is made to flow instead of solid objects being made to move. Silent wall switches in the home are a common example [Chapter 9].

Molten iron is agitated by induced eddy currents during steelmaking. Molten bauxite (aluminum oxide aluminum ore) conducts electricity during the electrolytic extraction of aluminum.

Another large class of liquid conductors is ionized solutions, such as salt water, in which conduction is by physical motion of the electrolyte's ions. In electrolytic conduction, the H^+ ions release hydrogen gas at the negative electrode. The gas may coat that electrode with fine bubbles that obstruct current flow; this effect is called *polarization*. Positive ions of metal may deposit as a metal film on the negative electrode; this is *electroplating*.

1.1.3 Gases

Ionized gases conduct electricity, with positive ions moving one way and free negative electrons moving the other way. This is what happens in neon signs, fluorescent lamps, and certain lasers. An ionized gas is called a *plasma*. Evaporated mercury is a gas much used as an electrical conductor, for example in fluorescent lamps.

In a sense, vacuum is a superconductor since it allows electrons to move through it with no voltage drop other than that used to accelerate the electrons. This occurs in television and other cathode-ray tubes and in other vacuum tubes. The motion of the electrons is accelerated by electric and magnetic fields and obeys the same Newtonian laws as do the motion of baseballs. (Gravity fields also accelerate the electrons but the effect is negligible in practical electron devices.) When electron speed approaches the speed of light, relativistic effects also occur.

In mass spectrographs, ions move through a vacuum, electro-statically accelerated and magnetically deflected. (Is a mass spectrograph a kind of curvilinear motor and therefore an EMD? If it is, so is every cathode-ray and other vacuum tube! Does it matter? "A rose by any other name would smell as sweet.")

Conduction in metals is by free electrons moving in the metal; conduction in an electrolyte is by ions moving in the solution; but conduction in a vacuum requires a separate source of electrons, such as a hot filament, or a separate source of ions.

1.1.4 Contacts

Conductor surfaces that touch one another conduct electricity from one to the other, depending on the conditions of the surfaces; the principal obstacles are the oxide and dirt layers that may be on the surfaces. Silver, gold, platinum, palladium, and rhodium—are called *noble metals* because they resist chemical attack such as oxidation in air. Therefore, they make good contact materials in switching devices even though their resistivities (other than silver's), and costs, are higher than those of copper or aluminum alloys [Chapter 9].

Carbon is also noble (although not usually called so), but its resistivity is high and it is difficult to make a low-resistance connection to it. (One way is to electro-plate the carbon with tin and then solder to the tin surface.) Carbon has the useful property that the resistance between carbon parts in contact with one another or with metal parts varies inversely with the force pressing the parts together. This is the basis for the following devices:

- The carbon granule microphone [Sections 7.4.1 and 12.2]
- A rheostat in which a stack of carbon plates is compressed by an adjusting screw
- A spot heater [Chapter 13]

Carbon does not melt, so it cannot be welded, intentionally or otherwise, and its oxides are gases that dissipate. These properties make it a valuable contact material in relays and circuit breakers [Chapter 9].

Silver is a special case. In air, it develops a sulfide coating (tarnish), but the coating is conductive. Since silver is cheap, it is a very common electrical contact material. A limitation of silver is its tendency to migrate over plastic surfaces and cause electrical leakage and short circuits.

In many switching devices, the contact surfaces are made to rub against each other *(wipe)* when they are brought together to scrape away dirt and chemical contaminants. This is particularly important for inexpensive switches and connectors whose contact surfaces are bronze or other non-noble metals [Chapter 9].

Many switching devices are enclosed in a vacuum or an inert gas to prevent atmospheric chemical or dirt contamination. Motion is transmitted to them via flexible diaphragms, bellows, magnetic fields, or gravity. Examples are reed switches, mercury switches, and vacuum circuit breakers [Chapter 9]. A vacuum or an inert gas also reduces arcing.

A solid conductor in contact with a liquid or gas is called an *electrode.* (Welding, medical, and chemical electrodes may also contact solids.) There is little problem with inadequate conductivity, but there may be severe problems with corrosion. In some instruments, the electro-chemical voltage between electrode and liquid would cause errors, so platinum electrodes may be used. In batteries, of course, these electro-chemical voltages provide electric power.

This *electrolytic corrosion* is sometimes combated by providing *sacrificial electrodes,* usually of zinc, which corrode, instead of the metal of the device to be protected, which is typically steel. Large devices, such as steel pipelines, are sometimes protected by electric voltages applied between the device and the earth. Even nominally dry contact between dissimilar metals can produce corroding voltages and currents if a return path exists. Corrosion of iron pipe in contact with copper tubing is a common example.

Electrolytic removal and deposition of metal are not always bad; this is the basis of electro-plating and electro-polishing and of a technique of metal purification. The original ampere-hour meter electro-plated silver from an anode to a cathode; weighing the electrodes gave an accurate measure of ampere-seconds, or coulombs. Another useful device is a running time meter comprising a small, sealed, glass tube with electrodes at each end and with a slug of metal at one electrode. A fixed current is sent through the device whenever the associated equipment is on. Transfer of metal to the other electrode is read on a scale calibrated in hours.

1.2 ELECTRICAL INSULATION

Insulation is as important as conduction; you must not only get electricity from here to there, but you must also prevent it from leaking out and spreading to where you do not want it.

The important electrical properties of insulating materials are dielectric strength, dielectric constant, dielectric loss, and arc and tracking resistance. For circuit breaker arc quenching, high thermal capacitance is of great value; hydrogen and sulfur hexafluoride are of particular value, as are solids that outgas *(ablate)* when heated by an arc [Chapter 9].

Dielectric loss is useful when it allows heating insulating material with high-frequency electric fields.

Insulating materials include solids, liquids, and gases. Most organic and inorganic compounds and non-metallic elements are insulators.

1.2.1 Solids

Many solid insulating materials are plastic laminates; and many others are polymers or mixtures, such as molding compounds with and without fillers. Coils are sometimes molded into solid plastic insulators, usually after being outgassed in a vacuum to allow the uncured plastic to penetrate the windings. Voids in electrical assemblies are sometimes filled with liquid *potting compounds* that harden in place. Solid insulation is fabricated by molding, machining, and laminating.

Organic solid insulators include molded thermoplastic and thermoset plastics, laminated plastic, wood, paper, cloth, and fiber.

Inorganic solid insulators include glasses, ceramics, mica, and glass filled with mica particles. Silicones are organic or inorganic, depending on one's chemistry semantics. They resemble organic compounds in many ways and make excellent insulation, particularly in place of other organics at high temperatures.

The important non-electrical properties of solid insulating materials are strength, stiffness, creep, brittleness, temperature resistance, moisture, chemical and fungus resistance, toxicity, fabrication properties, and aesthetic attractiveness [Group B].

1.2.2 Liquids

Liquid insulating materials are used to immerse and impregnate many transformers, capacitors, high-voltage cables, and high-voltage switches.

Long after they started to be used as insulators or manufacturing solvents, it was discovered that some materials were carcinogens, were poisons, damaged the ozone layer, or otherwise polluted the environment. Most of these materials were liquids, such as polychlorinated biphenyls (PCBs) and chlorofluorocarbons (CFCs), that spread easily or were volatile. Asbestos, a solid, was long used as an excellent heat-resistant insulation before it was discovered to be a carcinogen. (The fine strands of asbestos float through the air and hook into lung tissue.) [Chapter 27].

1.2.3 Gases

Gases, usually under pressure, are used instead of air as insulating materials in some devices such as large circuit breakers and high-voltage cables. Sulfur hexafluoride is a particularly good insulator. Hydrogen is an excellent electrical insulator and heat conductor, but it is a dangerous explosive.

Vacuum is an excellent insulator provided there is no source of either free electrons, such as a hot filament, or of ions; both can pass freely through it.

The properties of insulating materials are tabulated in the handbooks in [Group B].

1.3 ELECTRO-MAGNETISM

1.3.1 Magnetism

Many *electro-magnetic* devices are EMDs, but since the word *electro-magnetic* occurs in this book less frequently than *electro-mechanical,* it is spelled out each time. Alphabet soup is not a worthy end in itself.

A magnetic field is produced by every electric current. In most electro-magnetic devices, a current flows through a coil of many turns, N, the field of each turn adding to the fields of the other turns. The total magnetizing effect of a coil, *magnetomotive force,* (MMF), is measured as *ampere turns,* NI.

Magnetic material atoms have what amounts to electric currents in superconducting short-circuited rings. These may be thought of as aligned electron spins and orbits, although a physicist would call this description a rough approximation of quantum mechanical phenomena. In both soft magnetic materials, such as pure iron, and hard (permanent) magnetic materials, such as Alnico, MMF is required to change the directions of atom alignments from random to parallel and from parallel alignments in one direction to parallel alignments in the opposite direction. (It is as if there were friction among the atoms). The tendency for the alignments to remain in the same direction until forced to change by an MMF is called *hysteresis.* The difference between soft and hard *(permanent)* magnetic materials is the difference between low and high hysteresis. This difference is many orders of magnitude beween the least and the greatest.

Most materials have no effect on magnetic fields, which pass right through them, but some materials provide easier paths than air, that is, they have *permeability* greater than air. Principal among these are iron and nickel and their alloys. Parts made of these materials reduce the ampere turns necessary to produce a desired magnetic field. As moveable parts, they help convert electrical energy to mechanical work, as described later.

These *ferromagnetic* materials *saturate* at some level of magnetic field density. Therefore, they are not helpful for fields much above saturation density. Some iron-nickel alloys have very high permeability and very low hysteresis but saturate at a low magnetic field density. These alloys are useful in sensitive relays, certain transformers, and sensors for weak fields.

Some materials, such as bismuth, have permeability slightly *less* than that of air. They are *diamagnetic;* iron is *paramagnetic* or *ferromagnetic.*

Magnetic materials have *hysteresis;* their state of magnetization remains until reverse ampere turns demagnetize them. These reverse ampere turns may come from current in a coil or from an air gap. An air gap requires continual ampere turns to maintain continual flux and thus acts as such a current in a coil. Energy is dissipated as heat with each cycle of magnetizing and demagnetizing. Most AC devices use an alloy of iron and silicon, *silicon steel,* which has low hysteresis and high electrical resistivity, the latter reducing power dissipation from eddy currents. Materials with extremely high hysteresis become *permanent magnets* when magnetizing MMF is removed.

Another class of magnetic material is ferrite ceramics. They can be made with either very low or very high hysteresis, and they are insulators.

An extremely valuable use of magnetic hysteresis is in digital and analog memories. If you magnetize a piece of magnetic material having hysteresis, and then go away, and then return and examine the hysteresis state of the piece, and find that it is magnetized, you may say that the piece has *remembered* that you magnetized it. This is digital memory, which is used in digital computers [Chapter 10].

If you magnetize high-hysteresis material to less than saturation and later measure how much it was magnetized, this is analog memory, which is used in tape recorders for sound [Chapter 12]. (Sound can also be periodically sampled, the samples measured and the measurements digitized, and the digital measurements recorded or transmitted. This is what compact disc (CD) recording is and what modern telephone systems do for long-distance telephony.)

Since magnetic metals are electrically conductive, AC magnetic fields induce *eddy currents* in them that dissipate power in addition to the power dissipated by hysteresis. Therefore, the iron cores of motors, generators, and transformers are stacks of thin sheets *(laminations)* of magnetic steel. The laminations have insulation between them, typically an iron oxide layer on the lamination surfaces. The thinner the laminations, the less the eddy current loss. Hysteresis loss is unaffected by laminating. The magnetic steel used in most AC cores is an alloy of iron and silicon. It has less *core loss* (total eddy current and hysteresis loss) than other alloys or than pure iron. Ferrites are non-conducting and need no laminating but saturate at lower flux densities than do iron alloys.

Core-loss heating is used to heat iron parts for forging.

The force between permanent magnets is the reaction of their internal electric currents with the magnetic fields those currents generate. Electro-magnetic force is described below.

The properties of magnetic materials are tabulated in handbooks [B6, B9, B11, B13, B16].

1.3.2 Flux Linkage

To understand electro-magnetic force one must first understand *flux linkage*. Consider Fig. 1-1. Magnetic fields are represented by *lines of force* 1. The density of the field is represented by lines per unit area, and the direction of the field is represented by the direction of the lines. It works out that each line is an endless loop, so the total number of lines through any cross section of the entire field is the same. A bundle of lines is called a *magnetic flux*. The closed-loop path of a bundle of lines is called a *magnetic circuit*. (The array of all flux lines and the magnetic parts through which some pass is also called the magnetic circuit.) Where there is an iron core, as in a transformer or an electro-magnet 2,3 most of the lines 4 pass through the iron path. However some lines extend out and back through the envelope of the iron and are called *leakage flux* 5a, 5b.

There exist both approximate and rigorous mathematics and also finite element analysis for magnetic fields. The fields in both the iron and the leakage flux paths in devices made of iron 2,3 with air gaps 6a, 6b can

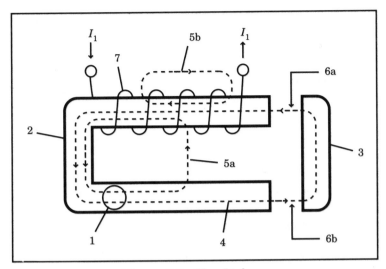

Figure 1-1   Flux Linkage

be calculated. However for our purposes we do not need the mathematics [A10].

An electric current flows in a closed loop in an electric circuit. An electric circuit is usually confined to the shape of the circuit conductors, and the conductors often include a coil of wire 7.

It has not been useful to coil magnetic circuits although there is no law of physics that says they cannot be coiled. An exception is the transformer core that is a spiral of magnetic strip instead of stacks of flat laminations. The flux crosses from turn to adjacent turn and the net effect is beneficial magnetically and economically.

A bundle of magnetic lines of force passing through—linking—coil turns of an electric circuit constitutes flux linkage. Flux linkage is an essential concept in understanding electrical and EM engineering.

One magnetic line linking a two-turn coil is equivalent in flux linkage to two magnetic lines linking a one-turn coil. Usually, some flux links some turns and some flux does not, as shown in Fig. 1-1. The total flux linkage is the sum, over all the turns, of the lines linking each turn. This is equivalent to saying that the flux linkage is the sum, over all the magnetic lines of force, of the number of turns linking each line.

There are few devices in which all the magnetic lines link all the circuit turns. Even in an iron-core transformer with no air gap, and certainly when there is an air gap 6a, 6b, there is some *leakage flux* 5a, 5b part of whose path is through the air. This leakage flux does not link all the turns. A perfectly made toroid, however, would have no leakage flux [Section 18.3].

In solenoids with large air gaps, it is the change of flux linkage with mechanical position that makes the device work, as we shall see.

In motors and generators, the flux crosses the air gap between stator and rotor, from tooth to tooth, with fringing around the edges of the teeth [Fig. 3-2]. Before finite element analysis (FEA) by computer, enormous amounts of effort were expended on approximate calculations of magnetic field shape. Reference [A10] describes the FEA approach and contains software to do it. New computer programs continually appear.

1.3.3 Induced Voltage

Voltage is induced in a coil by the rate of change of its flux linkage. This is equally true for huge turboalternators and for tiny instruments. The linking flux may come from any of the following:

1. An outside source, such as a permanent magnet moving relative to the coil

2. Electric current in another coil, such as in a transformer or a motor

3. Electric current in the coil itself, such as in an inductance

In each case, induced voltage is proportional to the rate of change of flux linkage.

If the flux comes from a permanent magnet or electro-magnet, we have either a *motor* or a *generator,* depending on which way power flows [Chapter 3].

If the flux comes from current in another coil, we have a *transformer* and the two coils have *mutual inductance.*

If the flux comes from current in the coil itself, we have an *induction coil* (or *inductor* or *choke*) with *self-inductance.*

If there is a movable piece of magnetic iron in the magnetic circuit, as in Fig. 1-1, when it moves it changes the shape and reluctance of the magnetic circuit. This changes the amount and shape of the magnetic flux and thus the flux linkage [Chapter 6].

In each case, voltage is *induced* by time varying flux linkage.

In AC devices, the flux linkage reverses cyclically, producing a reversing voltage, but the effect of geometry change is the same as for DC devices.

Flux cannot increase to infinity, so in DC motors and generators flux linkage is made to reverse; it is really AC. However, the coil connections are also reversed when the flux linkage starts to reverse, so the reversing voltage appears on the terminals as DC. Sliding brush *commutators* are EM switches for this purpose [Fig. 9-13]. *Commutation* is also done electronically: in generators by rectifiers and in motors by transistor switching circuits.

1.3.4 Electro-magnetic Force

In Fig. 1-1, if current I_1 is caused to flow in coil 7, flux 4, 5a, 5b is produced. Armature 3 is pulled toward core 2, the magnetic circuit reluctance is reduced, the flux linkage increases, and energy is drawn from the current source. *Motion in the direction of the force increases the total flux linkage of the system,* and the mechanical power comes from electric power to the coil. *The instantaneous electric power converted to mechanical power is current times the voltage induced by the increase in flux linkage. This is what happens in all electric motors and in all other electric actuators.*

We have assumed that current I is constant, which it is if a DC coil is exerting a force on either a stationary second DC coil or a stationary piece of iron. This is not the case while the device is in motion doing its work. While the system is in motion, changing flux linkage induces voltage that causes current change; thus the instantaneous force varies with time.

If DC voltage is suddenly applied to the coil, flux linkage builds up in the usual way for an inductance. If the rising force starts motion before steady-state current is reached, that motion begins with less force than if steady-state current were reached before the parts were released.

If AC voltage is applied to the coil, the action is more complicated. First, the current-time relationship depends on the electrical transient determined by the phase of the voltage when the switch is closed. Second, in an AC device, the current is affected only slightly by the coil resistance. It is the amount needed to generate such AC flux as will induce a back electro-motive force (EMF) equal to the applied EMF. As the iron moves, the time-varying magnetic circuit reluctance changes the amount of current needed.

There are three configurations in which electro-magnetic force occurs:

1. A single coil and iron. In Fig. 1-1, assume that the position of iron part 2 is mechanically fixed relative to the coil. A force appears between the two pieces of iron 2,3 that is proportional to I_1^2. If part 2 were not present, there would still be a force between iron part 3 and the coil that would be proportional to I_1^2, although much weaker.

2. Two coils, with or without iron cores. In Fig. 1-2 there are two coils 8,9 carrying I_2 and I_3, respectively. Some of the flux generated by each coil links the other coil. A force appears between the coils proportional to $I_1 \times I_2$. If one or both coils have iron cores to which they are mechanically bound, a larger force appears, but it still is proportional to $I_2 \times I_3$. If the two coils are connected in series so they carry the same current I_2, then the force is proportional to I_2^2.

3. A single coil and a permanent magnet: In Fig. 1-3 coil 10 carries current I_4 and is linked by flux from permanent magnet 11. A force proportional to I_4 appears between the coil and the magnet. Permanent mag-

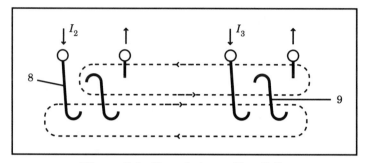

Figure 1-2   Force between Two Coils

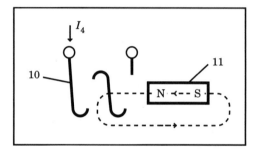

Figure 1-3   Force between Coil
and Magnet

net 11 acts like a coil with a constant current, as explained in the next
section.

The structure of Fig. 1-1 is used in the electro-magnets of most relays
and in lifting electro-magnets. The structure of Fig. 1-2, each coil having
an iron core, appears in DC shunt motors. The structure of Fig. 1-3 appears in loudspeaker voice coils, d'Arsonval galvanometers, and some electric motors.

Please note that in all cases *the force is proportional to the product of
two currents*. In case 1, both currents are the same, and the "product" is
a "square." In case 2, there are two distinct currents, although in the series
case, there is again a single current that appears twice. In case 3, the
permanent magnet produces a flux just as if it were a coil carrying a
current—which it really is, the current being electron spins and orbits.
In each case, the presence of soft iron changes the shape and increases
the magnitude of the magnetic field but does not change the "product of
two currents" law [Chapter 6].

1.3.5 Permanent Magnets

To repeat: Permanent magnets are pieces of matter in which many
electron spins and electron orbits have parallel axes so their individual

magnetic fields add up. The permanent magnet acts like a superconducting coil needing no external source of power to keep its electrons moving.

The force causing iron particles to attract each other in a magnetic field is the force on those electron currents. The attraction of particles to each other is the basis of the magnetic particle clutch and brake [Section 4.4.2].

1.3.6 Electric Power

If induced voltage causes current to flow in the same direction as the induced voltage, then electric power flows *out* of the coil, proportional to induced voltage times current, there is a generator. The amount of current depends on the impedance of the coil's circuit. The power is consumed by a load, for example, a utility customer's motor. The power may also be consumed by the resistance losses in the coil itself if it is short-circuited, which it is in some EMDs.

If *external* voltage causes current to flow in *its* direction, *opposite* to induced voltage, then electric power flows *into* the coil, proportional to external voltage times current; there is a motor or actuator. The law of the conservation of energy applies. Generator power comes from external mechanical force moving against $I \times I$ force, and motor power comes from $I \times I$ force moving against external mechanical force.

In most EMDs, there is energy transfer from electrical mode to mechanical mode or vice versa. In power devices, the power may be megawatts; in information devices, it may be microwatts.

More complete and mathematical explanations of electricity and magnetism are given in many textbooks, including references [A10, A11].

1.4 ELECTRO-MAGNETIC RADIATION

Electro-magnetic radiation extends in a continuous spectrum from extremely short gamma-ray waves to extremely long radio waves. Regardless of their wavelengths, all are combinations of interacting AC electric and magnetic fields.

The AC electric field induces the AC magnetic field that induces the AC electric field. The combination of fields, independent of the wavelength, moves forward in empty space at the speed of light. If this action is new to you and this description is hard to believe, you have my sympathy. The best I can do for you is to refer you to any formal textbook on electricity and magnetism, but it will say the same things much more rigorously, at much greater length, and with differential equations, wave diagrams, words like *displacement current, Poynting vector,* and *Maxwell's equations.*

To add to your troubles, the radiation always exists in discrete units called *quanta* or *photons,* each having a fixed amount of energy that varies inversely with its wavelength. There are sensors that can detect a single photon.

The good news is that, to understand the remainder of this book, you do not have to understand the last two paragraphs!

The spectrum of electro-magnetic radiation is usually divided into *frequency bands* named:

1. Gamma rays (shortest wavelength, highest frequency, greatest energy per photon)
2. X-rays
3. Visible light. (The longest wavelength is approximately twice the shortest wavelength. In musical terms, visible light spans approximately only one octave of the hundreds of octaves of known radiation.)
4. Infrared (IR) rays
5. Microwaves
6. Short radio waves
7. Long radio waves

Many of these bands have subbands with corresponding names.

In each band, EMDs are used for a variety of purposes, for example:

1.4.1 Gamma Rays

Gamma rays are used in inspection systems. Either a test body or a set of sensors is manipulated by an EMD with respect to a stationary beam of radiation. (A radioactive gamma-ray source, with its necessary shielding, is too heavy to move easily.)

1.4.2 X-Rays

X-ray tubes, film cassettes, objects being studied, and medical patients are all manipulated by EMDs.

1.4.3 Visible Light

Visible light is associated with a large number of EMDs. Light is *produced* by arcs (some controlled by EMDs), hot bodies (including incandescent lamps, fires, and the sun), lasers, fluorescent surfaces impacted by electrons or other invisible radiation, and phosphorescent surfaces energized by radiation.

Light is *directed and modified* by lenses, mirrors, prisms, shutters, and filters, all manipulated by EMDs.

Light *passes through* vacuum, most gases, many liquids, and some solids. Optical fibers are glass fibers, usually with glasses of different refractive index at different radii. They carry light signals for extremely wideband communication and are displacing plain wires and coaxial cables for long-distance communication.

Light is *sensed* by chemical deposits on photographic film, by photovoltaic and photoresistive cells, by emission of electrons from atoms impacted by photons in photomultipliers, and by silicon solid-state–charge coupled devices (CCD's). Examples include photoelectric cells of all kinds and cameras of all kinds, from vest pocket to the Hubble space telescope. All are manipulated by EMDs. Similar effects occur in video cameras that use CCD's as image sensors. Light is generated and used as a sensing and control medium in a variety of EMDs.

Light sources and sensors, such as searchlights, telescopes, and video cameras are aimed by EMDs [Section 8.1]. Many have photoelectric sensors in their control systems.

1.4.4 Infrared Rays

IR sources and mirrors are manipulated by EMDs. For example, the Sidewinder missile uses a spinning mirror in its IR target tracking system.

1.4.5 Microwaves

Microwave systems use an EM slotted line with a mechanically positioned sensor as an accurate measuring device for standing wave ratio. They use accurately manufactured waveguides, cavities, and antennas. Radar uses microwaves whose steerable antennas have diameters from a few inches to over one hundred feet [Section 8.1]. (But a *phased-array* antenna has no moving parts. It steers its beam by combining the radiation of many small, fixed antennas whose time phases are electronically varied.)

1.4.6 Radio Waves

Short- and long-wave radio transmitters use mechanically adjustable inductors and capacitors, some of quite large physical size. Some communication antennas for satellites and deep-space probes resemble large radar antennas. Some satellite and probe antennas are themselves EM unfurled and aimed.

1.5 NUCLEAR RADIATION

Radiation from nuclear fission comprises electro-magnetic radiation, electrons, and isotope particles (both ionized and non-ionized.)

Many EMDs are used with nuclear radiation, for example:

1. Control rod positioners and the *scram* electro-magnets that release control rods in nuclear reactors
2. Remote manipulators in *hot* rooms
3. Conveyors for mail or luggage in detectors for terrorist bombs and for foods being sterilized by nuclear radiation
4. Experimental portable detectors for land mines

Magnetic resonance imaging (MRI) medical diagnostic imaging uses no nuclear radiation at all but depends on the resonant frequencies of hydrogen nucleii (protons) oscillating in the patient's body. Large electromagnets and EM patient manipulators are parts of the system. (MRI, Magnetic Resonance Imaging, used to be called Nuclear Magnetic Resonance Imaging, NMRI. Medical terminology now omits the word *nuclear* because, after the Three Mile Island accident, many patients fear anything involving nuclear phenomena. The *nuclear medicine* profession suffered a major reduction in its practice.)

1.6 PIEZOELECTRIC AND MAGNETOSTRICTIVE EFFECTS

Another kind of electric voltage generation and force generation by electricity is *piezoelectric*. Certain crystals and ceramics, when stressed mechanically in certain directions, generate electric voltage between certain surfaces. Conversely, when electric voltage is applied to those surfaces, mechanical stresses and corresponding strains (displacements) are produced in the crystals and ceramics.

One of the most useful of these materials is quartz. When electroplated surfaces of a quartz crystal are connected to an appropriate electronic circuit, the crystal oscillates *mechanically* and establishes an extremely stable natural frequency for the circuit. Wristwatches, computers, and radios use such oscillators [Section 7.7.2].

Another particularly useful piezoelectric material is the ceramic, barium titanate. It is used in ultrasonic devices ranging from sonar to welders for plastics.

Magnetostriction is a magnetic dual of piezoelectricity. It is the shortening of certain materials in a magnetic field. Nickel is such a material. The effect has been used to adjust, with fine resolution, a massive centerless grinder [A34].

1.7 ELECTRO-STATIC EFFECTS

Electric charge can be stored without current flowing. The ability to store a charge is called *capacitance* and the devices to do so are *capacitors* [Section 18.2]. When charge is stored in a capacitor, electric voltage appears between its terminals.

Electro-static voltages are produced by friction between dissimilar insulating materials, by moving a body close to an existing electric charge *(electro-static induction),* and by connection to an electro-dynamic voltage source such as the generators described above that comprise AC voltage, a high-voltage transformer, and a rectifier.

Extremely high voltages are produced by mechanically spreading apart positive and negative charges. This occurs in nature when air currents charge clouds that then discharge as lightning; it occurs in the laboratory in EM machines such as the Van de Graff generator [Chapter 3] that uses a moving belt to spread apart charge, and it is produced by turning rotors of insulating material.

Electro-static voltage is used in EM dust precipitators, in xerographic photocopiers and laser printers, and for accelerating and steering electron beams in cathode-ray tubes and other vacuum tubes.

Electro-static motors can be made by applying voltage to electro-static generators. An electro-static motor has been made on a microminiature scale by etching shapes in semiconductor material. No coil winding is needed.

1.8 ELECTRO-EXPLOSIVE EFFECTS

Electric currents are used to ignite explosions by sparks in internal combustion engines and by hot wires in other explosives.

There is a large class of *powder-operated devices* in which a small explosive charge does useful work. Among such tasks are breaking a bolt in which the charge is embedded and moving a piston to drive a cable cutter. Many such devices are electrically detonated from a remote control or from an automatic circuit. Such devices are widely used to deploy parts of a satellite after its launch.

In industry, blasting explosives are detonated by a hand-powered generator to prevent accidental connection to a continuing electrical power source such as a battery.

The energy in an electrical discharge from a large capacitor is itself a kind of explosion. Lightning is one example, and the electric rail gun and the electro-magnetic forming machine are others [Chapters 13, 15]. Electro-magnetic pulses (EMP) from nuclear explosions are duplicated by such discharges to test the effects on electronic circuits.

Many EMDs are used in military safing, arming, and fusing devices [Chapter 15].

1.9 ELECTRO-THERMAL EFFECTS

Electric current generates heat in solids, liquids, and gases. The heating current may enter by conduction, or be driven by induction, or be displacement current between capacitor plates. Electric heating is produced by:

1. Current through resistance (solid, liquid, or gas)
2. Bombardment of arc electrodes by electrons or ions
3. Hysteresis in magnetic materials
4. Induced eddy currents in conducting materials
5. Dielectric loss in insulating materials

Heat from these effects is transferred to other bodies by conduction, convection, and radiation.

Electric ovens, spot and arc welders, electric furnaces, induction heaters, plastic welders, medical diathermy machines, and incandescent lamps use these effects.

Electric current also *extracts* heat in certain material junctions, providing a kind of refrigeration.

Conversely, heat generates electric voltage in junctions between different metals; such *thermocouples* are used as thermometers, measuring temperature indirectly by measuring voltage. Stacks of such junctions *(thermopiles)* are used as electric power generators.

Temperature affects electrical resistivity. This effect is used in resistance thermometers.

1.10 ELECTRO-CHEMICAL EFFECTS

EMDs, particularly sensors and sensor-controlled systems, use a variety of electro-chemical effects. The most commonly used sensor is the *pH electrode* that is measuring the acidity of a solution, but there are many other electrodes sensitized to other parameters.

Primary and secondary batteries power EMDs. Today's big hunt is for electric car batteries, primary or secondary, that are small, lightweight, capable of widely variable discharge rates, easily and quickly recharged with either electricity or fuel, non-polluting, safe, and cheap. The present art of electronic motor control makes the remainder of the electric car problem easy.

Electricity and chemistry are teamed in many processes including electro-plating and metal refining. Chemical plants use electrical transducers, control circuits, switchgear, and electrically powered pumps. Electrically powered proportional valves have not yet overcome the competition from pneumatically powered proportional valves, so many control systems are electro-pneumatic.

1.11 ELECTRO-BIOLOGICAL EFFECTS

Our bodies generate voltages which are sensed, recorded, and used by EMDs. For example, the *electro-cardiograph* is an EM recorder of the voltages that control the heart; the *electro-encephalograph* records brain activity; and the *electro-myograph* measures voltage signals in nerves.

Electric voltages applied to the body affect its processes in a variety of ways. For example:

1. The *cardiac pacemaker* generates voltages that control the heart. (Nominally, pacemakers have no moving parts, but the flexible conductors extending to the heart have severe EM problems.)
2. Electric voltages are used in pain-reduction therapy.
3. High frequency voltage produces *diathermy* (internal heating).
4. Electric current accelerates fracture healing.
5. Electric current cauterizes a cut blood vessel in surgery.
6. Shock: Some electric shocks *defibrillate* heart muscles, and others cure certain mental illnesses. Accidental electric shock, for example lightening, can kill. Dictators use electricity to torture their victims. You already know about the electric chair. Slaughterhouses electrocute incoming animals for painless death. Low-level shocks were once used as a mode of entertainment.

The patent office records many, many efforts to help the human body by applying voltage to it.

Lasers are used in diagnosis and surgery; EMDs manipulate lasers and their radiation.

X-rays are used for both diagnosis and therapy. EMDs are used to manipulate the X-ray tube, the film, the fluoroscope, and the patient.

Electric currents induced in the body by the magnetic fields from power lines and transformers are suspected of being carcinogenic. Designing these devices to minimize their external fields is an EM problem. Similar effects are suspected from the Navy's ELF (Extra-Low-Frequency) high power radio transmitter for communicating with submarines; its very long antenna is buried just below the surface of the ground.

Nerve voltages command EM prostheses.

Biological and medical laboratory instrumentation uses many kinds of EMD, such as automatic sample manipulators in analysis machines [Chapter 16].

1.12 ELECTRICAL-MECHANICAL ANALOGS
AND ANALOG COMPUTING

Many electrical and mechanical equations are mathematically identical. For example, the current produced by a constant voltage applied to an inductance has the same linear rise with time as the speed produced by a constant force applied to a mass. The two cases are *analogs*. A pair of analogs is sometimes called a pair of *duals;* inductance and inertia are duals. An internal combustion engine fueled by gasoline is a dual of an electric motor powered by a storage battery. A set of analogous parameters is as follows:

ELECTRICAL	**MECHANICAL**
Voltage	Force
Current	Speed
Charge	Displacement
Inductance	Inertia
Capacitance	Elasticity
Resistance	Viscosity (also friction, but friction is non-linear)
Time	Time
Power	Power
Energy	Energy
Frequency	Frequency
Strength (Dielectric)	Strength (tensile, compressive, shear)
Natural Frequency	Natural Frequency

Hysteresis exists in both fields. In electricity, it appears in magnetism; in mechanics, it appears in backlash and stiction.

An electric circuit can be designed to behave as an analog of a mechanical device, using the proper conversion constants and scaled with respect to time and magnitude. Thus a small, fast, electric circuit can quickly *simulate* the behavior of a large, slow, mechanical or thermal system. Such a circuit is a *direct analog computer.* The technique was once used but was replaced, first, by other forms of analog computer (below) and then by digital computers.

It is often useful to make a sketch or mental image of a direct analog computer to aid in thinking about a mechanism.

A more sophisticated form of electronic analog computer uses *operational amplifiers* to add, differentiate, and integrate. It uses EM servomultipliers and EM sine/cosine generators for trigonometric functions. Using the computer starts with writing the differential equations of the system to be simulated and continues with hooking up computing components to match the equations.

In the servo multiplier of Fig. 1-4, potentiometer 1 has brush 3 driven by servo 2. Output voltage X is proportional to input voltage Y times servo command voltage Z; $X = Y \times Z$.

In the sin/cos generator of Fig. 1-5, a special potentiometer comprises a square card 1 on which is a uniform resistance winding 2. Voltage V is applied to the winding. A set of four brushes is rotated by a servo whose input voltage is proportional to θ. The output voltage between brushes S1 and S2 is proportional to $V \sin \theta$, and the output voltage between brushes C1 and C2 is proportional to $V \cos \theta$. A two-phase synchro does the same thing with AC voltages.

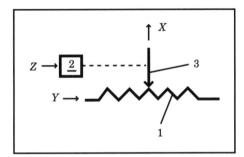

Figure 1-4   Servo Multiplier

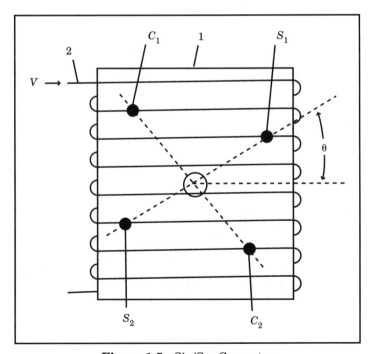

Figure 1-5   Sin/Cos Generator

Before electronic operational amplifiers were available, EM analog computers used integrating discs, Fig. 1-6. Disc 1 is driven by servo 2 at a speed commanded by input voltage Z. Disc 3 is driven by friction by disc 1 and is positioned radially by command input voltage Y, servo 9, lead screw 8, nut 7, and yoke 6. Disc 3 slides along output shaft 4 and is keyed to it so that shaft 4 rotates with disc 3. Shaft 4 turns transducer 5 that generates output voltage X.

$$X = \int Y \, dZ$$

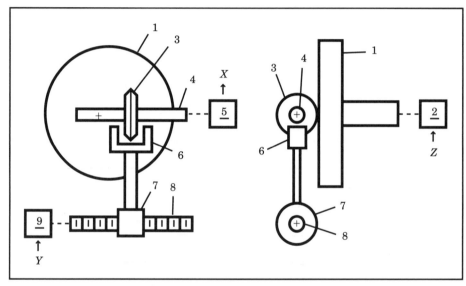

Figure 1-6   Integrating Disc

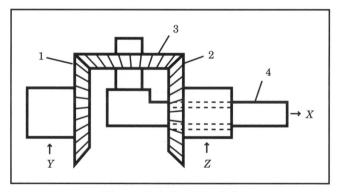

Figure 1-7   Differential

Addition and subtraction were done with gear differentials, Fig. 1-7, similar to a car differential. Output gear 3 is turned by a combination of input gears 1 and 2 and delivers its motion by shaft 4 extending through the hollow shaft of gear 2. Inspection of the figure will show that $X = \frac{1}{2}(Y + Z)$.

Non-analytic functions were stored on two- and three-dimensional cams or as relay-switched voltages.

There are pneumatic analog computers still used in industrial control systems in which air pressure and flow are used instead of electric voltage and current. They have the unique advantages that air pressure on a large diaphragm provides a cheap and reliable means to operate a control valve, and that pneumatic systems cannot trigger explosions in explosive environments.

Digital technology has replaced most analog computing with digital computing, so most EM computing devices are now obsolete. They are described here because the principles are interesting and may be revived, perhaps by yourself, in some new device.

1.13 DIGITAL COMPUTERS

There is a proliferation of databases, computing programs, and computer hardware useful in EM engineering. The programs include *finite element analysis* (FEA) programs for magnetic and electric fields [A10], heat flow and fluid flow fields, and mechanical stress fields. Formerly, except for simple cases, these were only roughly approximated, or subjected to very difficult mathematical analysis, or left to judgment and test.

Computer-aided drafting (CAD) is now commonplace. Many CAD programs include some mathematical calculations, justifying the translation of CAD to *computer-aided design* and the abbreviation CAE for *computer-aided engineering* [A36, A37].

Many EMDs are used in digital computers and are described later in this book. These include:

- Magnetic cards and readers
- Hard disk drives
- Floppy disk drives
- Tape drives
- Keyboards
- Switches
- Cables and connectors
- Printers
- Plotters

- Drawing digitizers
- Mouses (Mice?)
- Fans

Literally, "digital computers" include the ancient abacus (still used) and mechanical adding machines (obsolete). The first programmed digital computer was entirely mechanical, built in England in the nineteenth century by Professor Charles Babbage but not finished until recently [A21]. Experimental pneumatic digital computers with no moving parts have been built but have not been commercially successful. Pneumatic digital logic devices with moving part valves have some commercial success where their high temperature and radiation resistance and their safety in explosive environments give advantage over electrical systems. Pneumatically and hydraulically operated power valves provide logic control in fluid power systems. In present usage, *digital computers* means only electronic machines [Chapter 10].

1.14 COMMUNICATIONS

For many generations of development, until modern digital electronics came to the fore, all telegraphy and telephony were entirely EM. Even now, much EM engineering is used in the most modern communication systems.

Before Samuel F. B. Morse's invention of the telegraph in the 1840s, communication at a distance, *telecommunication*, used the following techniques, that need no detailed explanation here:

1. Human messengers, including postal systems
2. Carrier pigeons and dogs
3. Visuals (the first digital technology):
   - Semaphores [A52]
   - Smoke signals
   - Fires ("One if by land, two if by sea")
   - Rockets (one or more; different colors; still used with Very pistols)
   - Coded flags (still used by navies)
   - Hand signals (still used by infantries and by baseball catchers)
   - Heliographs
   - Blinking lights (still used by the Navy)
   - Others (for example, at Waterloo, Wellington galloped back and forth waving his hat to order the counterattack)
4. Sounds:
   - Bells, whistles, horns, sirens, buzzers, gunshots

- Speech via shouts, megaphones (now EM bullhorns), speaking tubes

5. Mechanisms:
   - Ropes, chains, shafts, push/pull rods, linkages. All are used to this day in railroad yards to move semaphores and rail switches from levers in control towers.

6. Hydraulics:
   - Analog displacement in hydraulic tubing

7. Pneumatics:
   - Message conveyor carriers and tubes
   - Analog pressure transmitted through tubing (still used in chemical plants and in some commercial building thermostats)

Morse's telegraph introduced instant long-distance electrical communication using digital codes. Early transmitters, relays, receivers, and printers were entirely EM. Most have gradually been superseded by electronics. The final receiver is now either a cathode-ray tube or an EM printer, and the initial device is an EM keyboard.

Telegraphy extended communication from the original person-to-person to include:

- Person-to-person *(telegraphy):*
  - E-mail consists of telegrams from personal computer to personal computer. It is much less expensive than earlier telegraph communication.
  - Facsimile (FAX) is telegraphy by raster scan instead of alphanumeric codes.
- Machine-to-person *(telemetry)*
- Person-to-machine *(remote control)*
- Machine-to-machine *(automation)*

Alexander Graham Bell, in the 1870s, extended digital telegraphy to analog telephony. Transmission of speech was then extended to transmission, first, of still pictures and then of moving pictures, *television.*

Telegraph and telephone signals were conducted first by electric current in open or twisted pair wire, then by coaxial cable or radio (some radio is via satellite and some is on microwave links), and now by light in optical fibers. For short distances, for example, amplifiers for theatergoers and for TV controllers, IR radiation is used.

The telephone originally used sound-modulated electric current. Now that analog technology has been replaced for long distances by digital technology, sound is sampled, digitized, and transmitted by what is essentially high-speed digital telegraphy [A22].

1.15 SOLID-STATE DEVICES

The first solid-state devices were rectifiers and photoelectric cells, with selenium as a principal material. (Photoelectric selenium is still the key material in photocopiers and laser printers.)

Marconi used a *coherer,* in which particles moved under the influence of radio frequency (RF) voltage, as his detector.

A galena crystal (lead sulfide) with a *cat's whisker* electrode was the detector diode in early radios.

The triode vacuum tube started the avalanche of electronic technology.

When the three-terminal transistor was invented, modern electronics exploded. Plates of pure silicon, *substrates,* have transistors, diodes, photodiodes, lasers, light emitting diodes, and integrated circuits built on them in two or more layers, using advanced photolithography. The manufacturing technology for solid-state devices is surveyed in [Sections 2.11.1 and 2.11.2].

Mechanically flexible solid-state devices are used as transducers [Chapter 7].

1.16 CIRCUITS, FIELDS, AND NON-LINEARITY

Many mechanical phenomena are much more complex and non-linear than most electrical phenomena. Many are three-dimensional *field effects,* such as the stress distribution in a non-uniform part, rather than two-dimensional *circuit effects.* Others, such as friction, are fundamentally complex. Therefore, many are described with only approximate mathematics and much pragmatism, while many electrical phenomena, being physically simpler, can be described by exact equations. Magnetic fields, particularly with portions in iron, electro-static fields, and electric current fields in conductors of complex shape, share the complexity of mechanical fields.

Fields bounded by certain geometric shapes are analytical. Among these are the electrical and magnetic fields around parallel cylindrical conductors, inside coaxial cables, and inside rectangular waveguides. Accurate mathematical descriptions of many mechanical, thermal, electric, and magnetic fields exist but are too complex, and therefore too expensive, for most engineering work. We have become fortunate in that computer programs now attack many such problems both by finite element analysis and by solving complex mathematical equations.

We have seen the distinction between electrical and mechanical phenomena begin to blur. Now we go on to primarily mechanical phenomena.

CHAPTER 2

Mechanical Engineering

2.1 SOUND

The acoustic spectrum varies from sounds too low in frequency to hear *(subaudio)* to sounds too high in frequency to hear *(ultrasonic)*. Sound waves are carried through gases, liquids and solids by waves of displacement and at speeds depending on the elastic stiffness and inertial density of the medium. Sound can travel through the earth or in the oceans for thousands of miles. Gases and liquids can carry only compression waves because they have no shear stiffness; solids can carry both compression and shear waves.

Some acoustic devices create sound, but not speech. Examples are bells, whistles, sirens, sonar transmitters, and musical instruments. Unwanted sound comes from noisy machinery and what other people think is music.

Speech is generated only by the human voice and by complex electronic circuits that emulate the human voice. Some animal sounds convey information and may be considered crude forms of speech.

Other acoustic devices reproduce sound, including speech, either simultaneously with its generation, for example, a speaking tube or a telephone, or by recording and playing back, for example, a tape recorder. Most record and playback systems use EMDs to produce relative motion between the record medium (tape or disc) and the recording and reading transducers (for example, magnetic heads, needle heads, laser heads.)

Sound vibration can modulate an electric current that then oscillates with the same waveform, is conducted over a distance electrically, and is converted back to sound, for example, by a telephone or radio.

Sound can be stored mechanically in phonograph grooves (on cylinders, disks, or belts), magnetically on wire, tape, or disc, optically on movie film or in digital/optical form, on CD discs.

Microphones, loudspeakers, sonar transducers, other ultrasonic transducers, phonograph record and tape recorders and players, and motion picture cameras and projectors are all EMDs.

The *Doppler effect* is the difference between received frequency and source frequency. It occurs when sound travels from a source moving toward or away from a receiver or when sound travels through a medium moving toward or away from a receiver. The speed of the sound *relative to the medium* is unaffected. The usual example is the received sound of a train whistle: The received pitch (frequency) is higher than the generated pitch when the train is approaching and lower than the generated pitch when the train is receding.

A reflecting surface creating an echo acts as a source with respect to the Doppler effect. The effect is used to measure the flow speed of a liquid in a pipe without inserting an instrument into the pipe. When the liquid is your blood and the pipe is your artery, you appreciate the benefit of this technique of flow metering.

(Electro-magnetic radiation also exhibits the Doppler effect. Police radar measures car speed, military radar identifies moving targets, and astronomers measure the receding speed of stars by the *red shift* in their spectra using EM telescopes.)

2.2 MECHANICAL PROPERTIES OF MATERIALS

Electrical properties of materials were discussed in Chapter 1. We now proceed to those mechanical properties of materials that affect the design and performance of EMDs. Among these properties are

- Strength
- Elasticity
- Hardness
- Ductility
- Corrosion resistance
- Fatigue

Reference [B16] has tables comparing engineering properties of different materials and providing bases for selection. It also gives typical applications for most engineering materials. It is the author's favorite source of material data when he is designing. There are many, many other sources of information about materials, including [Groups B, C, E, F, and G].

2.3 BEAMS

A *beam* may be a massive steel I-beam or a fine gold wire; the behavior of each is described by the same equations just as an electrical circuit consuming megawatts is described by the same equations as one consuming microwatts.

The mechanical properties of beams that affect the EMDs of which they are made include:

- Strength (in tension, compression, and shear)
- Stiffness (in tension, compression, and shear)
- Static deflection
- Dynamic oscillation

When a beam is bent by external forces, internal forces are produced that balance the external forces. (If these balancing internal forces were not produced, the external forces would accelerate the beam.) For example, the weight of a motionless diver bends a diving springboard, a *cantilever* beam. Figure 2-1 shows the internal forces in bent beam 1, comprising: tension 2 near those surfaces that bend into a more convex shape, compression 3 near those surfaces that bend into a more concave shape, and shear 4 distributed throughout. These internal forces are usually referred to as *stresses;* stress is force density, or force per unit area, typically in pounds per square inch.

Tension and compression are greatest at the surface and reduce to zero at the center, the *neutral axis*. Shear is zero at the top and bottom surfaces and is maximum at the neutral axis.

Distributed shear in a beam may be visualized by a mental experiment: Bend a deck of cards between your two hands. The cards slide over

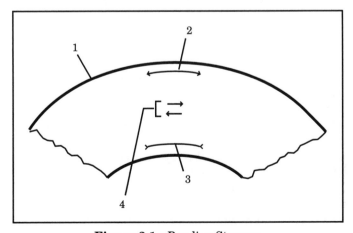

Figure 2-1   Bending Stresses

each other and you can see a small step at each end between each adjacent pair of cards. Now glue all the cards together into a single mass. Try to bend the deck again; it is much stiffer and the cards no longer slide over each other. The stress in the glue that prevents sliding between the cards is *shear.*

The *stiffness* of a beam is the ratio of force to deflection. The *strength* of a beam is the force required to break it. Be sure not to confuse *strength* and *stiffness.*

The strength and stiffness of a beam are quite sensitive to its dimensions. For example, in that diving board, the deflection under the stationary weight of the diver is proportional to the cube of the board's length and inversely proportional to the cube of the board's thickness. The strength, that is, the weight of diver that will break the board, is inversely proportional to the board's length and directly proportional to the square of the board's thickness.

Torque, such as the twist on a drive shaft, results in internal shear stress throughout the shaft, the greatest stress being at the surface and zero stress at the center. (To save weight, *torque tubes* are often used as shafts.)

Most springs are combinations of beams and shafts. (Helical springs are coiled shafts. Some are wound from tubing.)

All of these effects are subject to rather simple calculations for simple shapes and quite complex calculations for complex shapes. References [A7, A13, A20] are introductory texts, references [A14, A15, and A16] are engineering school texts, and reference [A19] is the standard advanced reference book for complex problems.

In all materials, tension stretches and compression shrinks. *Stress,* which is tension or compression force per unit cross sectional area, produces *strain,* which is stretch or shrinkage per unit length in the direction of the stress. Torsional stress and strain are analogous. This universal rule is put to work in *strain gauges* which are used to measure stress by measuring the strain it produces.

(The author knows of no instrument that directly measures *stress* in a part that is already under load, such as a beam in a bridge. In an electrical circuit you can measure electrical stress—voltage—anywhere at any time; what is the mechanical dual?)

EM *strain* gauges use resistive materials whose resistance varies with strain. Mechanical strain gauges lever up small motions to visible motions of a pointer on a dial. A brittle lacquer cracks in a pattern which indicates the surface strain. Models of transparent plastic show strain when illuminated with polarized light and viewed through a polarized light analyzer. There is an unusual strain gauge used to indicate the settling of the enormous, ancient church of Hagia Sophia in Istanbul. A rectangular pane of glass is cemented to a wall at the four corners of the pane. If and when the glass breaks, the fact of breakage warns of deformation and the shape of the break indicates the nature of the deformation.

Strain is used as an intermediate parameter in the diaphragms of pressure gauges, in measuring total force in *load cells,* and in measuring acceleration in *accelerometers.*

The entire subject of using physical phenomena to measure physical parameters is described in Chapter 7, "Transducers."

2.4 STATICS

It is necessary to understand the transmission and divisions of a force throughout a structure in order to design the structure. A *structure* is just as much the construction of a tiny instrument as the construction of an aircraft carrier; each is described by the same equations and each can fail if a part is overloaded. *(Fail* means either break or bend beyond useful shape. Manufacturers perform a great deal of intentional breaking and bending but with useful results.)

Statics deals with structures in which applied forces are not changing and the structure has reached elastic equilibrium. For example, the forces in the beams of a bridge are calculated by the equations of statics. (Actually, there are also *dynamic* forces in a bridge due to oscillation induced by wind. Since the Tacoma Narrows disaster in which a suspension bridge oscillated to destruction, these are now also calculated.)

The fundamental rules of statics are that if a part is not moving or rotating, all the forces on it must add up to zero and all the torques on it must also add up to zero. The basic approach to calculation is to resolve and sum the force and torque vectors at each joint in the structure. Some structures are *determinate* and their forces can be computed by simple vector algebra at all joints; others are *indeterminate* and require more complex calculations.

2.5 DYNAMICS

2.5.1 Newton's Laws

Newton's laws of motion apply equally and exactly to all bodies larger than an atomic nucleus, smaller than the solar system, and traveling much slower than the speed of light. Motor rotors and loads, circuit breaker contacts, and motor-driven inertia wheels used as orientation positioners in satellites are only three examples in which inertia and acceleration are essential parameters to consider.

The single equation

$$F = M \times A$$

(force equals mass times acceleration) sums up Newton's first two laws. Life is not always simple, however, and real-world problems may require applying this equation vectorially to each differential element of a system and integrating the effect over the whole system.

Newton's inverse square law of gravitation force simplifies on the earth's surface to a constant G times the mass of the object. Very small variations in G do occur due to non-spherical shape, different densities, and different altitudes of different portions of the earth. These variations are used in geological exploration and in computing the paths of long-range missiles. There are instruments that measure G with great accuracy [Chapter 7]. There is a device that actually *uses* the variation of G with altitude; it was used to torque satellites into vertical orientation.

It is important to realize that the force that accelerates a mass is the same kind of thing as the force that deforms a solid. Think about a sling-shot immediately before and immediately after you release it.

2.5.2 Gyroscopes

Modern gyroscopes are the ultimate in accuracy of all EMDs and all mechanical instruments. They are fundamental to EM guidance systems for surface vehicles, submarines, aircraft, and rocket-launched missiles and spacecraft. They are described in some detail in Chapter 7, "Transducers," but the basic physics is explained here, with Figure 2-2.

Imagine a stationary bowling ball resting on a frictionless spherical bearing such as described in Section 2.8.3. Assume a set of X, Y, Z coordinates as shown in the figure. Apply a momentary torque about the X-axis, Fig. 2-2(a). The ball acquires a spin about the X-axis. The torque was a vector **L**, it was applied for a differential time dt, and it produced an *impulse* **L** dt. The impulse resulted in an *angular momentum* vector **H** and by Newton's laws applied to rotation:

$$\textbf{momentum} = \textbf{impulse}$$
$$\mathbf{H} = \mathbf{L}\, dt$$

Next, suppose the bowling ball already had a spin about the Z-axis, $\mathbf{H}_0$, as in Fig. 2-2(b). If we apply the same impulse as before we add the same spin, **H**, as before. The added spin adds, vectorially, to $\mathbf{H}_0$, and the resultant spin is $\mathbf{H}_1$. By vector addition, the axis of $\mathbf{H}_1$ has *precessed* through angle p_1 from the Z-axis. A rotation is also a vector; $\mathbf{p}_1$ is the rotation vector of angle p_1, and it lies on the Y-axis.

Now suppose we do exactly the same thing, but this time, the initial spin is very, very fast, so $\mathbf{H}_0$ is very large, as in Fig.2-2(c). Now precession angle $\mathbf{p}_2$ is very, very small. In effect, the spin axis has remained almost unchanged, despite the impulse that has been applied. If we think of the spinning ball as having a human personality, it has resisted having its spin axis rotated, and if you hold a spinning gyroscope in your hands, that is just the impression you get.

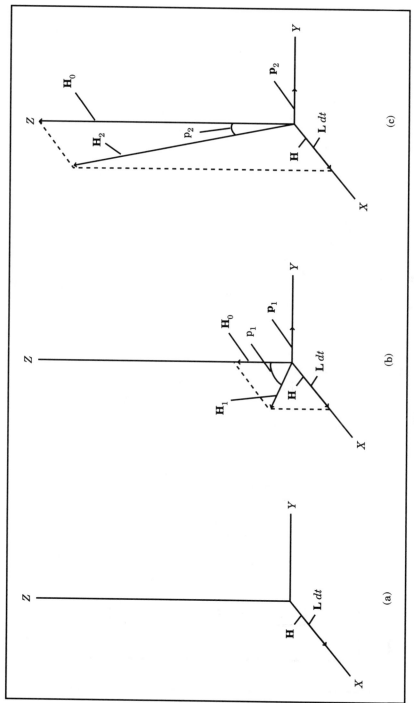

Figure 2-2 Precession

41

Now suppose we apply a series of differential impulses. Each impulse vector $\mathbf{L}\,dt$ adds another precession angle vector $\mathbf{p}_2$. In other words if we apply a continuous torque $\mathbf{L}$, the spin axis will continuously *precess* about the Y-axis.

To finish the job, in Fig. 2-3 let us replace the bowling ball and its spherical bearing with a heavy wheel 1 and axle 2 having the same moment of inertia as the ball and with the axle starting out along the Z-axis. Let us replace the spherical bearing with a frame *(gimbal)* 3 carrying the axle's bearings 4, and let the gimbal be supported on a second set of bearings 5 carried by outer gimbal 6 along the Y-axis. We now have a gyroscope as used in most inertial guidance systems.

If we rotate outer gimbal 6 around the X-axis, the same addition of impulses and momentums will cause gimbal 3 to precess around the Y-axis. It is this input rotation causing precession which enables the gyroscope to be used as an instrument to sense and measure the input rotation, as further explained in Chapter 7.

So far we have considered the *single-axis gyro.* If we mount gimbal 6 on a third set of bearings on the X-axis, we have a *two-axis gyro* capable of precessing around each of two axes.

Single-axis inertial guidance gyroscopes have, in addition, either electric torquers or springs and also angle measuring transducers on all gimbal axes, as further discussed in Chapter 7.

We have assumed that the angular momentum is stored in a spinning wheel. In fact, any oscillating body, such as a tuning fork, or a spinning

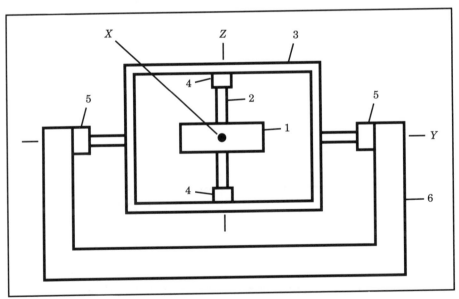

Figure 2-3 Gyroscope

fluid inside a hollow sphere also stores angular momentum and can be used as a gyroscope.

2.6 KINEMATICS

Kinematics is the study of the geometry of motion.

2.6.1 Moving Bodies

The kinds of moving bodies in EMDs include:

- Bodies rotating on the axes of bearings
- Bodies translating along linear guides or bearings
- Flexible cables, chains, or belts
- Gears:
    - Spur gears (parallel axes)
    - Bevel gears (axes intersecting at any angle)
    - Worm gears (axes at right angles, but not intersecting)
    - (Instrument gears (for information transmission)
    - Heavy gears (for power transmission)
- Flexible couplings
- Bearings
- Cams
- Screws:
    - Solid nut
    - Ball-bearing nut
    - Smooth shaft with helical roller:
        - Fixed helix angle
        - Variable helix angle

Kinematics also includes the study of:

2.6.2 Backlash

Backlash is looseness between parts transmitting motion such that, if the input motion is reversed, the input will move some distance before the output starts to follow. It most commonly occurs between gears and between a screw and nut. In mechanisms that move in only one direction, such as clocks, backlash may exist, but it never appears.

2.6.3 Preload

To prevent backlash it is common to add components, adjustments, and spring forces to maintain the same surfaces in contact despite a reversal of force or motion.

2.6.4 Hysteresis

Mechanical hysteresis is analogous to magnetic hysteresis. It is any effect in which a reversal of cause results in a delayed and lagging reversal of effect. Backlash is a form of hysteresis.

2.6.5 Coordinate Systems

In electrical circuits, other than waveguides, current flows only through conductors, and the geometry of the conductors does not matter except for inductive and capacitive coupling of nearby conductors. But in mechanical structures and motions (and in electric and magnetic *fields*), it is necessary to describe the geometry of shape, force, and motion as positions and orientations in three-dimensional space.

Regardless of the coordinate system used, it requires exactly six numbers *(six degrees of freedom)*, three linear and three rotary, to define the resultant of the forces and torques on a body and the position, velocity, and acceleration of that body. Several coordinate systems are used: orthographic, polar, airplane, and fluid displacement.

2.6.5.1 Orthographic

This is the system of Cartesian analytic geometry. Position is described as X, Y, and Z coordinates, and orientation is described as vector components parallel to the X-, Y-, and Z-axes.

At one time, mathematics was part of a curriculum that included Latin and Greek, so we have words like *orthographic* instead of *right angle picture*, its meaning.

2.6.5.2 Polar

In spherical coordinates, position is described as a radius from an origin and two angles that define the orientation of that radius. Orientation is defined as vector components along any convenient axes.

Cylindrical coordinates are a combination system in which the position radius is located as one angle and one linear distance from the origin to the inner end of the radius.

2.6.5.3 Airplane

A very useful nomenclature for orthographic coordinates is *airplane coordinates*. The six degrees of freedom are

1. Position north-south
2. Position east-west
3. Position up-down
4. Orientation in pitch
5. Orientation in yaw
6. Orientation in roll

Position, orientation, and motion can be easily described and visualized with these images, even in a telephone conversation. Sign of position is intrinsic: North is plus and south is minus, east is plus and west is minus, up is plus and down is minus. Sign of orientation is described: pitch up or down, yaw left or right, roll left or right. "North-south" is usually taken as a system axis rather than as true geographical north-south. This nomenclature will be used from here on in this book.

2.6.5.4 Fluid displacement
Pneumatic and hydraulic displacement is by volume or mass along a conduit and requires no geometric axes any more than does electric current in wires. (Free flow in open channels is different but does not occur in most EM engineering.)

2.7 FRICTION

Both static and sliding friction occur in all moving part devices. The phenomena are complex and the elementary physics book laws are only approximate. Design requires testing and must allow for variations in the coefficients of friction due to temperature, wear, lubrication, corrosion, speed, load, and load distribution.

As a first approximation, friction force is proportional to normal force and is independent of contact area. The independence from area is plausible because if the area increases, the force per unit area decreases, the friction per unit area decreases, and the total friction remains the same.

Starting friction, or *static friction*, is the force required to get sliding started. *Sliding friction*, or *running friction*, is the force to maintain motion and is always less than starting friction. Starting friction is often called *stiction*. Sliding friction force is independent of sliding speed; it is named *Coulomb friction*.

There may be alternating starting and sliding friction between two bodies in relative motion if there is elasticity between part of a body and its friction surface. The two portions of that body may oscillate with respect to each other. Alternate sliding and static friction may be described as *chattering*.

Lubrication reduces friction, but its laws are complex.

2.7.1 Starting and Sliding

We know from our physics courses and from practical experience that static friction is high and prevents any motion at all until a large force breaks it loose. We also know that, once static friction is overcome, sliding friction requires only a smaller force to maintain motion. We tend to assume that both forces must be in the direction of the desired motion; this is not so. Once sliding starts in any direction, static friction is overcome in all directions. (Remember what happens to a car when it skids on ice.)

A useful application of this principle is in machines that must insert a close-fitting cylinder into a close-fitting hole. If the cylinder is continually rotated during the insertion, it will go in easily if it is free to self-align. This can be done by supporting it at the end of a pair of universal joints that transmit torque but do not constrain transverse position or orientation. A short piece of rubber tubing does the same thing.

An elegant linkage called a remote center compliance device (RCC) was once developed at MIT to do the same thing without rotation. One of the Detroit automation companies bought a license and put on a show to demonstrate it. The visitor was invited to insert a ball bearing into a close-fitting hole by hand; he always failed and the bearing always jammed. The RCC then did the job in one stroke. When the author's turn came, he rotated the bearing as he inserted it and it went right in. He showed the technique to the exhibitors who threw him out.

Actually, you already know and use this principle. What do you do when you withdraw a cork from a champagne bottle? (If you have not done so before, merely rotate the cork; the gas pressure will push it out.)

We have been speaking of friction as an enemy, although we know that it is also a friend; without it everything would skid around and screw fasteners wouldn't hold.

2.7.2 Vibration and Friction

Vibration helps overcome friction by reducing normal force each half cycle, thus reducing the force needed to overcome friction.

An elegant application of vibration to overcoming friction is in vibratory feeders for granular material. The direction of vibration is approximately 45° from the vertical. The vertical component alternately reduces and increases friction, and the horizontal component moves the material horizontally during the high-friction half cycle and moves only the trough back during the low-friction half cycle.

Part feeders for automatic machines do the same thing, usually with helical motion. The first time you see a stream of parts moving *upward* along a helical track with nothing to drive them but vibration is a startling experience.

In precision servomechanisms, starting friction is sometimes overcome by adding a small, oscillating, fictitious error signal to keep the motor in a state of motion so that it can respond to a small real-error signal without having to overcome starting friction. The oscillation is called *dither.*

2.7.3 Vacuum and Friction

There is a physics class demonstration in which a block slides down an inclined plane, first in air and then in vacuum; the coefficient of friction in vacuum is much higher than in air. In air, a thin film of air sticks to surfaces and is a lubricant. In the space age, this effect is important for mechanical design of spacecraft; all rubbing and rolling surfaces must be lubricated and the lubricant must not evaporate in vacuum, especially if it might condense onto solar cells and instruments.

2.7.4 Tribology

The science of friction has been studied intensively in recent years under the name *tribology*. Material combinations, galling, and lubrication are among the subjects studied. This book does not cover the field but if you find yourself hemmed in with a difficult problem, this is the word to use in seeking help.

2.8 BEARINGS

Most EMDs have moving parts, and the motion of these parts is guided by rotary or linear bearings. The function of bearings is to guide with minimum friction and maximum accuracy. A variety of bearing types have been developed, as follows.

2.8.1 Plain Journal Bearing

Sliding bearings are concentric cylinders, cones, or flanges. Cylinders provide rotation about an axis, cones provide radial looseness adjustment, and flanges provide axial constraint.

Figure 2-4(a) shows a plain journal bearing. The shaft (*journal*) 1 rotates in *bushing* or *bearing* 2 installed in stationary member 3 and is usually lubricated. Lubricants include oil, water, grease, soap, graphite, and molybdenum disulphide. The bushing material provides lower friction and less wear than if the shaft turned directly in member 3, whose material may be chosen for other reasons than friction and wear.

Most shafts are steel or stainless steel for maximum strength in minimum diameter; the smaller the diameter, the less the friction torque.

There are many bushing materials and details of design, depending, in part, on the available lubrication. Bronze is common. Sintered bronze powder impregnated with oil provides lubrication without other lubricating means being provided. Graphite and plastics such as nylon and teflon have low friction and wear even without lubrication and are usually used without lubrication.

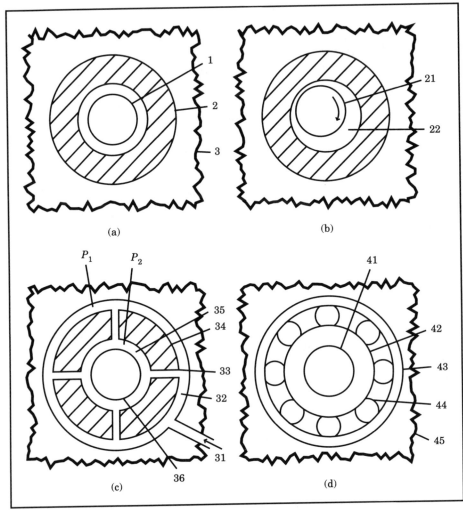

Figure 2-4 Journal Bearings

2.8.2 Hydrodynamic Bearings

Figure 2-4(b) shows a bearing in which journal 21 rotates continuously in an oil bath. Friction between journal and oil drags the oil 22 down under the journal so that the journal rides on oil and is not supported by metal to metal contact. The result is low friction and wear in this *hydrodynamic* bearing.

Hydrodynamic bearings are configured for radial loads, as shown, and also for axial loads. Axial hydrodynamic bearings are used for the thrust bearings of ships' propeller shafts and for the support bearings of water turbine generators. Some high-speed instruments such as gyroscopes use hydrodynamic bearings with air as the lubricant.

The magnetic heads on computer hard disk drives ride on films of air dragged under the heads by the speeding disks; these are also hydrodynamic bearings.

Continuous high speed is required for fluid pumping in hydrodynamic lubrication, so it is used for continuous motion and not for intermittent or reversing motion.

2.8.3 Hydrostatic Bearings

Figure 2-4(c) shows a bearing in which air or oil 31 is pumped into annular reservoir 32 at pressure P_1. A number of restricted passages 33 through bearing 34 carry the lubricant to the narrow clearance 35 around journal 36, from which it leaks out endwise. Pressure drop through passages 33 makes pressure P_2 in clearance 35 less than source pressure P_1.

If shaft 36 becomes eccentric in its clearance, the clearance reduces at one portion and increases at the diametrically opposite portion. At the reduced portion, flow decreases, pressure drop in its neighboring passages decreases, and clearance pressure increases. At the opposite portion, the opposite occurs and clearance pressure decreases. The difference in clearance pressures forces the journal back to the center. Extremely accurate centering with great resistance to being forced off center is achieved.

One extreme example are the hydrostatic oil bearings that support the 200-inch telescope on Mount Palomar; at the other extreme are the hydrostatic air bearings between granite blocks in accurate linear slides and in some shaft bearings. Hydrostatic air bearings have gaps of only a few microinches and have high load-carrying capacity.

A spherical hydrostatic air bearing is used to support a spinning, free-sphere gyroscope. A spherical hydrostatic air bearing is used to support a stable platform for inertial guidance instruments.

If you are an electrical engineer, you will get more insight into this device if you draw an analog in which pressure is voltage and the radial, circumferential, and axial passages are resistors.

Birds, airplanes, gliders, and helicopters are also supported by air pressure under their wings and rotor blades, the air being pumped downward by wings and rotor blades *(airfoils)* moving as inclined vanes through the air. By Newton's third law, the force accelerating air downward equals the force supporting the wings and blades upward, balancing the force of gravity.

2.8.4 Rolling Bearings

Figure 2-4(d) is a rolling bearing. The bearing comprises *inner race* 42, *outer race 43,* and balls or rollers 44. Shaft 41 fits tightly in the inner race, and the outer race fits tightly in stationary member 45. The races have ridges to prevent the balls or rollers from escaping. Rolling bearings have little friction at any speed, including starting.

So far we have seen bearings whose purpose is to withstand radial load. Analogous structures are used in *thrust bearings* to withstand axial loads and in combination bearings to withstand both radial and axial loads.

2.8.5 Conical Pivots

Figure 2-5 shows a conical pivot in which conical journal 1 fits conical bearing 2. This construction is used only in very small sizes in instruments such as voltmeters. The journal is turned to a fine point, and the bearing is a polished jewel. The small rolling radius results in low friction.

These descriptions have been basic. Innumerable details and variations exist for a variety of reasons. There are engineers whose entire careers are devoted to rolling bearings [A1, A13, A14, A25, B1, B2, B3, B4, B5].

2.8.6 Knife Edge and Flexure Bearings

Figure 2-6 shows a number of bearings having certain benefits where rotation can be limited. Figure 2-6(a) shows a knife-edge bearing in which sharp edge 1 of hardened steel rolls on flat 2 of hard stone. The center of rotation is the point of contact, and its location is accurately maintained by parts not shown. Such bearings are used in laboratory analytic balances, and crude versions are used in other beam balances. The stone is sometimes made as a blunt V to position the knife-edge; it may also be of metal. Friction is extremely low, and stiction is zero.

Figure 2-6(b) is also a rolling edge bearing used in the electro-magnets of relays and contactors. Magnetic armature 21 has an edge that is constrained by means not shown into the corner formed by electro-magnet core 22 and flat member 23. The magnetic reluctance between core and armature is minimal, and the overall cost is low.

Figure 2-6(c) is a *flexure pivot*. Thin, flexible strip 31, the *flexure,* is clamped to members 32 and 33 leaving a short, unclamped portion between. Friction and stiction are zero, but there is some finite stiffness in

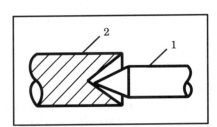

Figure 2-5   Conical Pivot

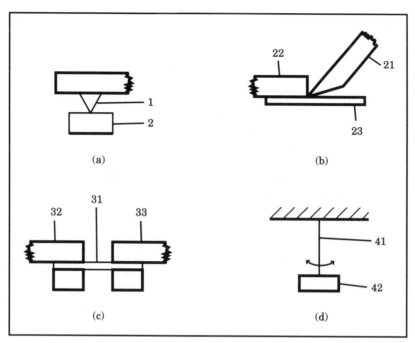

Figure 2-6 Limited Rotation Bearings

the flexure. Among flexure uses are for relay armatures, pendulum suspensions, and vibratory feeder suspensions. Flexures sometimes serve as flexible electrical conductors as well as bearings.

Another kind of flexure is a wire, quartz fiber, or narrow ribbon twisting about its axis, Fig. 2-6(d). Twist flexures have no transverse stiffness and so are used with hanging loads or are put under tension to support horizontal loads. Here again friction and stiction are zero. The torsional stiffness of the fiber is often used as the restoring force in the instrument.

The variations and refinements are legion [A14, B1, B3, B4].

2.8.7 Linear Bearings

Most linear bearings use the same principles as rotary bearings: Flat surface slides on flat surface, either dry or lubricated. Lubrication is either by oil, oil or air pumped between the surfaces, a plastic bearing strip between the surfaces, or with a set of rollers or balls between the surfaces. There is no simple analog to the hardened precision races of rotary ball or roller bearings. Track grooves are machined into the slides and sometimes furnished with inserted hardened wires on which the balls roll. Otherwise, the load per ball must be reduced to the stress capacity of the unhardened groove.

Figure 2-7 shows a recirculating ball circuit. Member 1 moves on fixed member 2 rolling on balls 3. It is necessary to keep feeding balls into the leading end and retrieving them from the trailing end. The recirculating path between member 1 and path enclosure 4 provides such feeding and retrieving.

(There is a basic difference between rollers and wheels. A ball or roller carries load between opposite tangents. At one tangent the ball or roller moves at load speed, at the other tangent it is stationary, so the center moves at half load speed and recirculation is necessary. A wheel, on the other hand, carries load between a central axle moving at load speed and a single tangent moving at zero speed. The wheel moves at load speed, and the top of the wheel moves through the air at twice load speed. No recirculation is needed, but an axle and its bearings must be provided.)

Figure 2-8 shows a use of recirculating ball circuits. Moving member 1 rolls on fixed member 2 consisting of a hard steel shaft. *Ball bushing* 3 contains a number of recirculating ball circuits 4.

Using the ball bushings of Fig. 2-8, the moving member must roll on two parallel shafts to constrain against roll and usually requires two spaced ball bushings on one of the shafts to constrain against pitch and yaw. Furthermore, a load between a convex ball and a convex shaft causes high stresses at the point of contact.

Figure 2-9 shows a variation in which the fixed member 2 has more than one grooved path for the balls. This both increases load capacity and provides roll constraint so a single bar can be used instead of two bars.

2.8.8 Ball Screws

A third variation is the ball screw, Fig. 2-10. Axial interference between screw 1 and nut 2 is via balls 3 rolling in matching helical grooves in both. Path 4 recirculates the balls. Ball screws have very little friction compared to screws with meshed solid threads and have much greater load-bearing capacity for the same diameter.

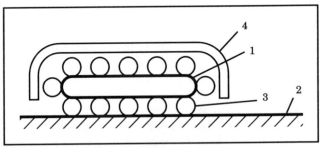

Figure 2-7 Recirculating Ball Circuit

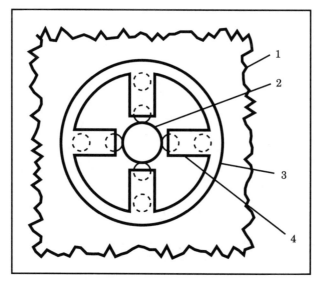

Figure 2-8   Ball Bushing

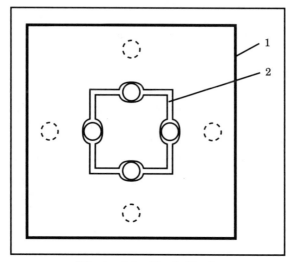

Figure 2-9   Single Track

Large quantities of rolling bearings, of great variety, are made to extreme accuracy by specialized manufacturers. All the rolling bearings described above are standard commercial components available in wide ranges of size and varieties of features.

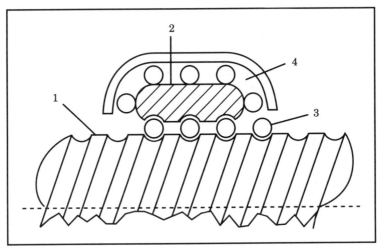

Figure 2-10   Ball Screw

2.8.9 Rolls

A roll is a curved portion of one part that rolls along another part. The part motion is a combination of rotation and translation. A roll is a portion of a part and provides for less than a full revolution; a roller is a separate part and may rotate without limit.

2.8.10 Buoyancy

On most bearings, most of the loads is the gravity weight of the supported body. If the body is partly or totally immersed in liquid, the buoyancy force reduces or totally cancels the gravity force. Friction is reduced both by the smaller diameter of rotary bearings permitted and by the reduced load on both rotary and linear bearings. Examples:

1. The weight of a boat is exactly balanced by its buoyancy; no bearings are needed at all; and only viscous drag, wind, water current, and wave-generation force resist motion. Starting friction is zero. That is why ships, including river and canal barges, can carry so much more load than trucks.

2. Balloons, blimps, and dirigibles are buoyant in air and therefore require no power to hold them up, as do airplanes and helicopters. It was once hoped that dirigibles would become commercial freight and passenger carriers, but storms and burning hydrogen (as in the *Hindenburg* disaster) stopped the effort.

3. Most inertial guidance gyroscopes are sealed in a cylinder immersed in oil with neutral buoyancy. In Fig. 2-3, gimbal 3 is the cylinder. The precession bearings 5 are merely fine wires in jewel bearings. Friction is extremely small.

4. Permanent magnetic force is equivalent to buoyant force and is used to reduce bearing friction in watt-hour meters and other instruments.

5. Electro-magnetic force levitation (MAGLEV) is a technique to reduce friction in experimental high-speed trains.

6. Electro-static force levitation is used in certain inertial guidance instruments. Both MAGLEV and electro-static levitation are unstable and require feedback control of the levitating force to maintain the position of the supported load.

2.9 HEAT

A full knowledge of thermodynamics in not needed to understand or design most EMDs, but a knowledge at the college physics level is necessary to understand many. For example, many EMDs generate heat, either intentionally or incidentally, and many, such as bimetal actuators, respond to heat usefully. In electrical packaging, heat dissipation is a fundamental task of the design.

Heat is the source of most of the power of most of our society. Both internal combustion engines and external combustion engines (for example, steam turbines) are *heat engines*. Nuclear power becomes heat before it becomes electricity. Water power comes from solar heat evaporating surface water that returns down by gravity.

2.10 FLUID POWER

Pneumatic and hydraulic pumps, cylinders, valves, motors, and plumbing are commonly used in EM systems. Fluid statics and dynamics at the college physics level is essential knowledge.

Electro-hydraulic and electro-pneumatic valves of many kinds, both switching and throttling, and a variety of pressure, temperature, flow, and position transducers are the principal EMDs in fluid power systems. Most fluid power systems are powered by electric motors.

2.11 MECHANICAL FABRICATION

An EMD is an assembly of components designed and fabricated for the device and also of standard components listed in commercial catalogs. Although fabrication methods are not "phenomena" in the scientific sense, they are technology, and there are rules governing design just as scientific rules govern behavior.

Most fabrications are either:

- Castings
- Moldings
- Stampings
- Machinings
- Weldments

or combinations of them.

There are many less common fabricating processes that can yield highly advantageous parts. Among these are:

- Photoetching and chemical milling
- 3D photolithography
- Laser cutting and welding
- Electrical discharge machining
- Ultrasonic machining

Many fabrications are made of semifinished materials such as tubing, extrusions, and rolled sections.

2.11.1 Manufacturing Solid-State Devices

Rigid solid-state devices are made by the following processes:

- Photolithography [Section 2.11.2]
- Electro-plating
- Vacuum deposition
- Etching
- Laser trimming
- Diffusion
- Ion implantation
- Thermocompression bonding of lead wires

2.11.2 Lithography

Lithography is a very old art that has evolved to a state and functions its originators never dreamed of. Originally, it was a fine-art technique for producing prints from hand-drawn art and, in part, it still is. It was based on the different adhesion of ink to oily and to dry areas on a flat stone, whence the name. The root, "lith," comes from the Latin for *stone,* as in "monolith," a single piece of stone.

Photolithography copies an optical image onto the stone by photo-chemistry, without human artistry, but the "stone" is now a metal plate or a semiconductor disc. First a *resist* is coated all over, then image light hardens portions of the layer, and finally the unhardened portions are dissolved off. Most multiple-copy printing today is by photolithography, usually with a rubber roller transferring—or *offsetting*—the ink pattern from plate to paper; the usual word for the process is *photo-offset.*

A version of photolithography is *photoetching.* Instead of producing areas in which ink either adheres or does not adhere, it generates chemical resist patterns. Where the resist is dissolved off, chemicals etch out shallow pockets in the metal plate. Originally, and still, the pockets hold ink pressed into the pockets. Sometimes the pockets form a visible image and are used as nameplates or decorations.

Photoetching all the way through thin metal sheets is a burr-free alternative to die stamping. *Printed circuit* patterns are photoetched in the copper layers of copper-plastic laminates. A similar process called *chemical milling* is chemically etching deep shapes in thick metal. (Sculptors have chiseled shallow or deep shapes, including lettering, into stone surfaces for thousands of years; their work is called *relief sculpture.* Shallow grooves in metal plates made by small hand gouges is called *engraving. Photoengraving* is photoetching.)

Integrated circuits are made by alternately depositing a resist, performing an additive or subtractive operation through it, dissolving off the resist, depositing the next resist, and so on. As pattern lines and spaces became smaller and smaller, the diffraction of visible light limited the fineness of the pattern, so light is being replaced by X-rays and even electron beams.

This has been a quite superficial description of one of the most sophisticated manufacturing technologies in history.

Xerography, another form of photolithography, is photoelectrostatic rather than photochemical. A selenium surface is coated with a uniform electro-static charge and then exposed to an image. The surface is a kind of giant photoelectric cell. Illuminated areas lose their charge to the underlying metal drum. Dry ink powder, called *toner,* is then applied to the surface and adheres electro-statically to the areas *not* discharged. The surface is then pressed against plain paper which receives the toner. Heat melts the toner and it bonds to the paper. Xerography is the process of most of the vast amount of photocopying done today. If the optical image is produced with a spot of laser light which is switched on and off as it scans the surface, the device is a *laser printer.*

If the final paper—or plastic film or metal plate—is coated with a light-sensitive chemical, the process is either photography or *blueprinting.* The original blueprinting chemistry is obsolete, replaced by the *ozalid* process, but the word survives.

There are textbooks on manufacturing processes [A17, B5, B18], but the advertising in mechanical design magazines, backed by manufacturers' literature, is an excellent mode of continuing education. Chapter 13 has additional material on manufacturing.

2.12 FLEXIBLE DEVICES

2.12.1 Silicon Devices

Silicon is no more rigid than stone or steel. A thin piece, a *chip,* may be coupled to external forces that deform it elastically, and strain gauge resistors built onto the chip measure its strain. Typically, the chip is mechanically coupled to the metal diaphragm of a pressure transducer, but with low-pressure fluids that do not attack silicon chemically, the chip itself may be the diaphragm. Signal-conditioning electronics are deposited either on the same chip as the strain gauge resistors, or on a ceramic substrate to which the chip is bonded and connected, or on a separate chip mounted on a common substrate. Two of the benefits of such solid-state pressure transducers is that they can be made very small and rather cheaply.

The same photolithography process that etches out pockets in the substrate can also cut all the way through. It produces contoured chips with cantilever shapes like diving boards and tuning forks. When such a shape is gripped at the root of the cantilever and accelerated, either in a single direction or in oscillation, the cantilever bends and the bend is sensed by strain gauge resistors. Thus accelerometers are made for inertial navigation and to release automobile air bags. Mechanical stops to limit deflection under shock are added by silicon diffusion bonding.

2.12.2 Quartz Devices

What can be done in etching and electro-plating silicon can also be done with quartz. Furthermore, quartz is piezoelectric, so strain generates voltage without the use of strain-sensitive resistors. Quartz has zero mechanical hysteresis. Tiny quartz accelerometers are made which are so sensitive they can measure variations in the earth's gravity [Chapter 7].

2.12.3 Other Devices

Photolithography and chemical etching have produced tiny electrostatic motors, actuators, valves and orifices for pneumatic and hydraulic circuits, piezoelectric oscillating mirrors, and relay actuators. Fiber-optic elements coupling optical fibers in different ways are made with similar techniques.

2.13 COMMERCIAL MECHANICAL COMPONENTS

As in the electrical art, in the mechanical art there is an enormous variety of standard components. A few examples are fasteners, rubber bumpers, shock absorbers, seals, and bearings.

The formal approach to finding what you want is to guess the existence of a category, look it up in a purchasing directory such as *Thomas Register* [E1], and contact manufacturers. However you will not look for a component whose existence you do not know. Furthermore, your knowledge of available components will stimulate your ideas in both electrical and mechanical design. Reference [A8, Part 2, "Designing with Commercial Components"] has a systematic classification of most common components.

Regular scanning of design magazine advertising is a good way to grow your knowledge. Some are listed in the references [Group D. Magazines with New Product Announcements].

- Reference [Group F. Mail Order "Department Store" Catalogs] lists catalogs of both common and uncommon mechanical components.
- Reference [Group E. Directories] is the starting point for finding component manufacturers whose catalogs will teach you about their components. The author's preference is *Thomas Register* [E1].

2.14 NULL MEASUREMENT

Many EMDs are measuring instruments, so a comparison of displacement measurement and null measurement is appropriate.

Displacement measurement is done by instruments which displace in proportion to the quantity being measured. For example, a galvanometer moves its pointer across its scale, against the restoring force of its spring, in proportion to the current in its coil; a caliper's user displaces its slide along its scale to match the dimension being measured; the weight on a spring scale stretches the spring and moves a pointer along a scale. The accuracy of the instrument depends on the calibration of its scale, on calibration of any restoring spring in the system, on friction, and often on temperature and other effects.

Null measurement balances the instrument's stored standard parameters against the unknown parameter until a simple displacement instrument measuring the *difference* reads *zero*. Accuracy is independent of the accuracy of the displacement instrument, whose accuracy at zero is sensitive only to its hysteresis and whose range is unrelated to the quantity being measured. Accuracy depends only on the accuracy of the built-in standard parameters. It is common for null instruments to be accurate to four significant figures. The range of a null instrument can be much larger than the range of a displacement instrument.

A typical null instrument is a laboratory weight balance in which the weight of the unknown is compared to the weight of a group of standard weights. (Instead of standard weights, a modern version uses the force between two coils whose currents can be accurately measured.) Most gyros and accelerometers in inertial guidance systems, the most accurate mechanical instruments made, use electro-magnetic force balance systems.

An electrical null instrument is a Wheatstone bridge that measures electrical resistance by comparing an unknown resistance to the sum of a set of known resistances.

2.15 OSCILLATION

2.15.1 General

Oscillation is the back-and-forth transfer of energy between two different forms. Examples:

1. The balance wheel of a mechanical watch stores kinetic energy when it is in motion, and the hair spring stores potential energy as mechanical stress and strain. When the wheel speed is maximum, the spring stress is zero, and all the energy is in the kinetic form. When the wheel speed is zero, the spring stress is maximum, and all the energy is in the spring in potential form. In between, either the spring is accelerating the wheel and transferring energy to it or the wheel is bending the spring and transferring energy to it.

2. Bells, chimes, drums, and violin strings all resemble that balance wheel except that the usual amplitude is too small to see, the motion may have harmonics, the motion may be different in different portions of the object, the inertial member and the elastic member are the same member, and the vibration is at audio frequency.

3. The air in an organ pipe or flute is like a violin string except that the inertial element is the mass of air, the elastic element is the elastic stiffness of air, and the oscillating displacement is longitudinal instead of transverse.

4. An electrical capacitor and an electrical inductor connected in a closed loop transfers electro-static energy from the capacitor dielectric to magnetic energy in the space around the coil and back again.

In each case, the oscillating device *resonates* at its *natural frequency* unless it is driven at some other frequency.

Wave motion is traveling oscillation, that is, radiation. Examples:

1. In sound waves, the inertial element is the mass of the medium and the elastic element is the stiffness of the medium. The medium may be solid, liquid, or gas. Sound is mechanical oscillation.

2. In electro-magnetic waves, electro-static field is one energy store and electro-magnetic field is the other energy store. As the electro-static field in one region changes, it induces an electro-magnetic field in an adjoining region, and as the electro-magnetic field changes, it induces an electro-static field in an adjoining region.

Almost all oscillations lose some energy during each cycle. In some cases, the loss is into heat through some form of friction. In some cases, oscillation energy is radiated, such as sound from a musical instrument or light from a hot filament. Electro-magnetic radiation in a vacuum is an exception, so light from a distant star reaches us reduced only by its spreading out laterally and by losses in the diffuse matter in space through which it passes. The words *dissipation, attenuation* and *damping* are used for such losses. Materials that dissipate oscillation energy are called *lossy*.

To start and to maintain an oscillation, power must be provided. Musical instruments are provided with power from compressed air (from lung or compressor), friction (mostly from bows), pluck-and-release, impact, or (in an electric organ) an EM-driven diaphragm. In the cases of pluck-and-release and impact *(percussion)*, a fixed quantity of energy is provided instead of steady power.

2.15.2 Feedback Control

Some oscillation is desirable, such as the sound from a violin. Some is undesirable, such as the shaking of an airplane rudder moved by an unstable feedback control or the squeal in an audio system having too much feedback. All feedback control systems are supplied with power, and most physical devices have some natural frequency at which they can oscillate. A major part of feedback control engineering is preventing oscillation.

The reverse problem of feedback control is to generate sustained oscillation when such oscillation is desired. Examples are audio oscillation in electric organs and radio frequency oscillation in radio transmitters.

The most accurate and stable common *electrical* oscillators use a *mechanical* resonant device—a quartz crystal—to establish the frequency. The crystal oscillates like a bell. (Actually, the *most* accurate use molecular oscillation, the Maser being an example.) Quartz is piezoelectric: When the crystal bends, it generates a voltage between metallic areas on its surface, and when a voltage is applied to those areas, it makes the quartz bend. The metallic areas couple the crystal to an electrical circuit that receives and transmits those voltages. Mechanical resonators such as quartz crystals are more stable and have a higher Q than purely electrical oscillators (Q is the ratio of the energy stored in an oscillating portion to the energy dissipated per half cycle oscillation. High Q means low loss.). [Section 7.7.2, "Clocks and Timers"].

2.16 MECHANISM

Mechanism is the application of kinematics, statics, and dynamics to real devices. The art has developed over thousands of years, so the present rate of development does not compare with that of electronics.

Mechanisms transfer input motions to output motions. (In common usage, "mechanism" may include actuators and output loads and may be entire machines.) The variety of mechanisms is large so this book has left out all but a cursory description, a few examples, and some philosophical comments about mechanisms as portions of EMDs [B4].

A major type of mechanism is the *linkage,* which is a number of rigid members coupled by rotary and linear bearings. The linkages most exposed to public view are in earth-moving machines manipulating earth buckets.

A second type of mechanism is rotary and linear gear arrays and the looser variations of rotary and linear chain, belt, and friction couplings.

A third type of mechanism uses two- and three-dimensional cams as fixed-function memories.

Computing mechanisms using all of the above types were used in fire control and general-purpose analog computers before they were displaced by electronic digital computers. They are still used in mechanical watches.

Mechanical calculators combined digital and analog processes. They were digital in that they worked in the decimal system and used gear ratios between decimal places. They were analog in that instead of using binary states to perform binary arithmetic, they used ten analog increments of displacement to represent the ten steps in decimal arithmetic. Electronic circuitry cannot easily do this and so electronic digital computers are constrained to binary arithmetic and need decimal-to-binary and binary-to-decimal converter circuits to communicate with people.

(Analog and digital are not totally separate worlds. An early numerical control system positioned a jig borer to 0.0001 inch in 6 feet, using a combination system.)

Some mechanisms can be represented by one or two parameters, a pair of gears for example, but others have complex two- or three-dimensional geometries. Input to output transfer functions for both two- and three-dimensional mechanisms can be quite complex. In comparison, most electrical circuit elements have only one or two parameters (perhaps non-linear). However, they can be economically compounded into physically small assemblies such as printed circuits and integrated circuits of enormous component count, component density, interconnection complexity, and high speed. Equivalent mechanisms would be monsters of enormous size, cost, and slowness.

Elaborate mechanisms are used in EMDs in circuit breakers and simple ones in door latches. The possibilities for future developments in EMDs incorporating mechanisms are quite large.

PART **II**

Understanding the Devices and Systems

The next set of chapters describes electro-mechanical devices, systems, and fields of application. In them you will find the remarkable diversity of the EM art.

A few of the examples are obsolete, typically replaced by all-electronic devices, and one may predict that some others will become so. (This is an article of faith among some engineers.) They are described here anyway, not just for the insights of history, but because they are interesting and because their principles may be revived in some new device, perhaps by you.

CHAPTER 3

Generators and Motors

3.1 GENERAL

Motors are the most numerous of all EMDs. In cars, they start engines and adjust seats and mirrors. In homes, they blow air, mix food, and clean carpets. In factories, they move everything that moves. You might take a moment and see how far you can extend this list, just off the top of your head.

Generators are far less numerous than motors; on average each generator supplies power to many motors. Utility generators are among the largest machines ever built. However, generators and motors are so similar in their principles that they should be explained together.

Generators and motors are the only EMDs taught in depth in engineering schools, and there is a great deal of depth to teach. They are subject to both shallow and very deep mathematical analysis. This chapter is a superficial review for those who have already studied the subject fully and is a qualitative first introduction for those who have not. Detailed mathematical treatment can be found in many textbooks including references [A10, A11, A31, A32].

Generators convert mechanical power to electrical power and motors convert electrical power to mechanical power. The difference between a generator and a motor is the direction of power flow. In a generator, mechanical power enters the shaft, overcomes electromagnetic torque, and is converted to electric power coming out of the windings. In a motor, electric power enters the windings and is converted to mechanical power that electromagnetic torque puts out through the shaft. The structures of some kinds of motors and generators differ very little whereas the structures are quite different in other kinds.

As stated above, motors resemble generators except that power flows in the reverse direction; that is, electric power flows in and mechanical power flows out. Some motor loads store energy, either kinetic or potential or both. To brake such a load, the motor is made to change role and act as a generator: Power flows in reverse from mechanical to electrical and returns to the electric supply wires. If this power is dissipated in resistors, the system is said to have *dynamic braking;* if the power is returned to the electrical system for use elsewhere, the system is said to have *regenerative braking.* For example, mountain railroads use regenerative braking, partly to recycle the energy instead of dissipating it and partly to avoid the wear and heating of friction brakes.

All motors and all generators have a stationary member *(stator),* a concentric rotating member *(rotor),* and a magnetic field crossing the *air gap* between them. The magnetic field rotates with respect to either the rotor or the stator or both. *Electric currents in both the stator and the rotor generate the magnetic field and interact with it to produce torque.*

The current in either the stator or the rotor may be electron spins and orbits instead of current in a winding. These electrons may be in permanent magnet atoms or in soft magnetic steel atoms.

A typical configuration is shown in Fig. 3-1. Cylindrical frame 1 is mounted by feet 5. End bells 2 carry shaft 4 on bearings 3. Magnetic stator 6 having windings 7 is fixed to frame 1. Magnetic rotor 8 having windings 9 is fixed to shaft 4.

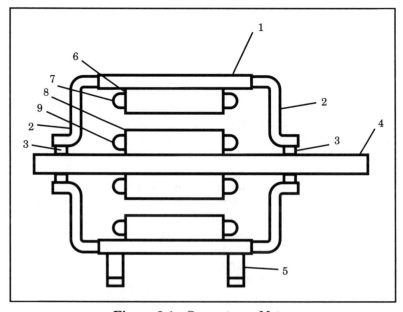

Figure 3-1 Generator or Motor

There are innumerable other configurations. The rotor may be outside the stator, mounting may be a flange on an end bell, the rotor may have no winding at all, there may or may not be a commutator, the stator may be fed by a complex electronic circuit, power may be AC or DC, and so forth. However, there are only a few basic operating principles for all the varieties.

Linear motors are equivalent to rotating motors that have been unrolled. An important difference is that the length of the moving portion is the minimum providing the required number of poles, while the fixed portion may be of unlimited length. The result is a low-inertia system.

Figure 3-2 shows how both sets of windings 7,9 are usually embedded in *slots* 13 in the stator 6 and rotor 8. The parts between the slots are called *teeth* 14. Tooth and slot construction is a compromise between having minimum reluctance in the magnetic circuit, that is, all tooth, and having minimum resistance in the winding, that is, all slot. The unfortunate consequence is the distortion of the magnetic field, particularly as rotor and stator teeth and slots pass each other. Harmonic voltages, currents, torques, and acoustic noises appear, and the efficiency and other characteristics of the machine are affected. The full mathematical description of the machine is quite complex.

The problem is reduced by skewing the teeth and slots into reverse helixes so there is no sudden change as tooth edge passes tooth edge (or salient pole edge); however, this alone does not solve all the problems.

In DC generators and motors, Fig. 3-3, a segmented copper commutator 10 is mounted on the shaft with its segments wired to the rotor windings. The commutator reverses each winding as the commutator's segments pass stationary, sliding, carbon brushes 1 that connect the com-

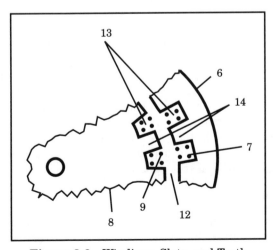

Figure 3-2   Windings, Slots, and Teeth

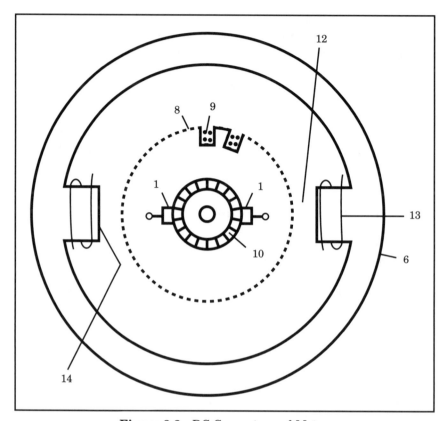

Figure 3-3   DC Generator and Motor

mutator to the external circuit. Thus the machines are AC internally and DC externally.

A *commutator* is a copper ring cut into segments insulated from each other. It is mounted on the rotor with the armature windings, and the segments are connected to those windings. A pair of stationary carbon brushes 1 slide on the segments and connect them in sequence to the output terminals. While each winding has an alternating voltage generated in it as it rotates past the north and south poles of the field, its connections to the terminals are simultaneously reversed by the commutator and brushes. What comes out is direct current.

In some machines, there are solid slip rings that connect to rotor windings independently of rotor position.

3.1.1 Speed

A discussion of speed, torques, and power is appropriate at this point. Generators are designed to match their prime movers. Most water turbines, other than Pelton wheels, turn slowly. Therefore, their generators are quite large physically and have many poles to generate standard 60 hertz (Hz) voltage at the slow speed.

On the other hand, steam turbines can be made to operate quite fast. Steam turbine-powered alternators run at 3600 RPM, which is the fastest an AC generator *(alternator)* can run and generate 60 Hz voltage. (The turbines could be made to run much faster.)

When weight is of paramount importance, as in airplanes, the usual frequency is 400 Hz. Generators and motors can run at 400 Hz $\times$ 60 seconds = 24,000 RPM and be much smaller and lighter than 60 Hz machines which can run no faster than 60 $\times$ 60 = 3600 RPM. (Series motors can run faster, but they have the limitations of their commutators and of speed variation with load.) Similarly, 400 Hz transformers are also much lighter than their 60 Hz counterparts.

Mechanical power is proportional to speed times torque, so for the same power, torque is less as speed is more. In all machinery, for a given power, the size, weight, and cost vary directly with torque and therefore inversely with speed. Similarly, in generators and motors, for a given power, the size, weight, and cost vary directly with torque and therefore inversely with speed. The price of low torque and high speed, however, includes high hysteresis and eddy current loss and the necessity to gear fast motors to slow loads.

Most generators operate at approximately a fixed speed and in only one direction. Utility generators must operate at exactly the designated speed to be operated in parallel with other utility generators. On the other hand, automobile generators must perform usefully over the full range of engine speed; DC output and auxiliary voltage regulators are provided to enable this to happen.

Motors have many more speed requirements than generators. Some loads can operate at a fixed speed but many cannot. For example:

1. Railroad trains have high inertia and stiction and require high torque during starting and less torque when running. Their motors must work at speeds from zero to maximum.

2. Conveyors and hoists must operate at different speeds.

3. Servo systems for vehicle steering, machine tool motion, and so forth, require exactly controlled speed over a wide range and in both directions.

3.2 DC GENERATORS

The *armature* of a DC generator is the rotor in which voltage is induced and which supplies power to the load. The *field* is the other cylinder that produces a DC magnetic field either by DC in a winding or by permanent magnets. In Fig. 3-3, the DC generator has a stationary field 6 and a rotating armature 8. AC voltage is induced in the armature and rectified to DC by switching commutator 10. The field is usually a set of cantileved, or *salient,* poles 14 with their inner ends in cylindrical form and energized by DC windings 13.

Instead of a soft iron field carrying a magnetizing winding, a permanent magnet field may be used.

Automobile generators use a rotating permanent magnet field inducing AC in a stationary armature and solid-state rectifiers to convert the AC to DC. This configuration eliminates the cost, maintenance, and unreliability of the sliding brush commutator. The windmill generators of proximity fuses do the same.

A special case is the homopolar generator in which the armature is a solid copper cylinder or disc with slip rings instead of a commutator. It was Faraday's experimental generator and motor. Commercially it provides very heavy currents at very low voltage.

Instead of a cylindrical field and armature, disc shapes are sometimes used, with the magnetic field being axial instead of radial.

All electrostatic generators are DC. They start with electrostatic charges produced by friction and mechanically move the charges apart with either rotating discs or moving belts (Van de Graff generator); the increased separation produces high voltage.

3.3 AC GENERATORS

The usual AC generator has a rotating field winding and a stationary armature winding. This is permissible because no rectifying commutator is needed, and is desirable because armature connections need not pass through sliding brushes.

In larger sizes, the field has a winding provided with DC from an auxiliary DC generator via slip rings; sometimes the auxiliary generator is on the same shaft as the main generator. In small AC generators, permanent magnet fields are used.

AC generators are usually three-phase because of the benefits of three-phase in power transmission and in motors. (The operation and benefits of three-phase power are outside the scope of this book but are treated in all introductory texts on electrical engineering.) For special purposes, generators may be single phase, two-phase, or multiphase.

An induction generator has no field windings other than a short-circuited rotor just like an induction motor (see below). When it is connected to a system containing a synchronous generator and is driven by a prime mover above synchronous speed, it transmits power from the prime mover into the system.

3.4 MAGNETOS

A magneto is a special generator for producing high-voltage pulses for engine ignition. It comprises a permanent magnet rotor and a stator having two coils. The first coil has a small number of turns and is short-circuited by a contact. The contact is opened by a cam on the rotor shaft at the angle where there is the greatest rate of change of flux through the stator. Until the contact is opened, current is induced in the first coil and bucks change in stator flux. When the contact opens, this current is interrupted and stator flux rises rapidly. This rapidly rising flux induces the high voltage in the second coil that produces the ignition spark.

3.5 ALEXANDERSON GENERATORS

In the early days of radio, special AC generators were built with very many poles. The generated voltages had frequencies in the low RF range, provided high power, and were used for radio telegraphy. Alexanderson generators were made obsolete by vacuum tube oscillators.

3.6 MOTORS

Motors are the duals of generators; they receive electric power and put out mechanical power. Every rotating electric motor has a rotor and a stator, mostly concentric cylinders as shown in Figs. 3-1 and 3-2. One cylinder carries an armature winding connected to the source of electric power. Except in DC brush motors, the armature is the stator 6 so that the electrical connections can be direct and not pass through sliding brushes. In DC brush motors, the armature is the rotor and has a commutator and brushes to convert DC line voltage to AC voltage at the armature winding. These are the duals of DC generators in which the commutator converts AC armature voltage to DC line voltage.

In AC motors, there are two or more armature windings displaced from each other around the armature and they carry alternating currents in different time phases. The result is a rotating magnetic field that torques the rotor around. This is the case with every kind of rotating electric motor.

DC brush motors carry AC in their armature windings, produced by commutator switching. The magnetic field produced by the armature windings rotates in the opposite direction from the rotating armature, so it remains stationary in space while the rotor turns.

A variety of effects are used to produce torque between the armature's rotating magnetic field and the other cylinder. These are described below.

In a DC brush motor, the armature is wound on the rotor as in Fig. 3-3. The current is converted to AC in each of several armature windings by a commutator and brushes which act as a reversing switch for each winding in turn. Thus the current in each winding is AC of approximately square wave form. So-called brushless DC motors are really AC motors with electronically fed stator armatures and with permanent magnet rotors. They may include transducers to tell the electronic power supply control circuits their angular position. Despite the cost, complexity, and maintenance requirements of the commutator, the DC brush motor is the most controllable in speed and torque in all sizes.

All motors are really AC because their armatures carry reversing currents, although *AC motor* usually refers to a motor fed from AC utility lines without electronic modification of the power.

All motors except DC brush motors have armature windings on their stators. These windings produce the rotating magnetic field which torques the rotor around.

There are several kinds of rotor:

1. One kind of rotor is a winding on an iron core supplied with DC current, producing a field with a set of magnetic poles fixed to the core. This field locks to the rotating armature field and rotates with it; this is a synchronous motor. Wound-rotor synchronous motors are made only in large sizes; in small sizes, permanent magnet rotors are used. In a stepping motor the rotating field rotates in small steps under electronic control.

2. In a DC brush motor with a rotating armature and commutator, the armature field counterrotates because of commutator switching, so the armature field is stationary in space. The armature field locks to the stationary stator field, which is produced either by a DC winding or by permanent magnets.

3. A permanent magnet. Everything is the same as in 1 except that a permanent magnet replaces the electromagnet. Permanent magnet motor fields are made of either ferrites or metal alloys. The latter are more costly but are stronger and have made possible major reductions in motor size and weight.

4. Stepping motors. Most stepping motors have many teeth in each pole of the rotor. They produce a small angular displacement from each electrical step, that is, from each step in armature field rotation. These interact with a different number of teeth on the stator. Some stepping motors are divided into two magnetic portions whose teeth are shifted with respect to each other. They have common armature windings, as if there were two motors on the same shaft with a common armature and a common shaft. Most stepping motors are driven by electronic power supplies.

5. A short-circuited winding. In AC motors, the rotating field generated by the armature induces currents in this winding, and the induced currents react with the armature field to produce torque. At equilibrium, the rotor RPM is slightly less than the magnetic field RPM. This *slip speed* between the rotor and the armature field results in rotor currents of slip frequency. This is the *induction motor* invented by Nikola Tesla and is the most common type of motor in use. (Tesla had a patent on the induction motor. He licensed it to George Westinghouse, but he was later persuaded to abandon his royalties, which would have brought him enormous wealth. An incredible man, his biography is well worth reading.) [A55].

The short-circuited winding is usually a set of aluminum bars embedded in slots in the iron rotor and connected to aluminum rings at each end. The bars and rings are usually a single casting with the iron rotor embedded in it. This winding is called a *squirrel cage.*

6. Some large induction motors have a wire-wound rotor connected through slip rings to adjustable external resistors. The larger the resistances, the greater the slip.

7. A rotor may be made of high-hysteresis magnetic material. Hysteresis lag produces starting torque, and finally the motor reaches synchronous speed with the high-hysteresis material acting as a permanent magnet.

As you may suppose, there are many variations on these basic themes; some are described below.

3.7 ELECTRONIC POWER SUPPLIES

Armature voltages may be from the utility line that provides polyphase, AC, sinusoidal voltage, or they may be synthesized by one of several kinds of electronic power supply.

Electronic power supplies and their controlled currents have enabled development of a new generation of motors. These include stepping motors, variable-speed motors, high-speed motors, and servomotors. The development of sophisticated, reliable, and economical electronic control has revolutionized motor and motor application technology. Electronic power supply currents may have complex waveforms and timing and obey a variety of rules.

Electronic power sources for the armature provide the following means of control. Common to many of the schemes is a transducer on the motor shaft to tell the control the angular position of the rotor.

Variable-voltage DC provides variable speed for DC brush motors. A DC motor's speed is approximately proportional to its armature voltage, regardless of load, if its field strength is fixed.

At one time, a separate motor-generator set provided variable DC voltage under the control of a potentiometer in the generator's field; this is the classic *Ward-Leonard drive.* Now a small and inexpensive controllable rectifier does the same thing. The controller may also provide dynamic or regenerative braking. (Note that the separate motor-generator set is an electro-mechanical amplifier.)

An electronic variable-frequency power supply varies the speed of an AC induction or synchronous motor down almost to zero speed. The voltage is varied directly with the frequency to maintain a field strength independent of frequency. The maximum speed of an AC motor with a 60 Hz power supply is 3600 RPM, but an electronic power supply can provide any frequency and therefore can drive a motor at very high speed.

For driving motors in servomechanisms, the DC or AC electronic motor controller provides reversible current and can provide dynamic or regenerative braking.

For servomechanisms with digital input and digital feedback, the electronic controller responds to the net digital pulses and transmits an appropriate, reversible, analog voltage to the motor.

If the motor is a stepping motor, the controller energizes the armature windings in discrete steps corresponding to the digital *input* pulses rather than with an analog voltage. No feedback is needed with stepping motors except to verify the correctness of the operation. For microstepping, the power supply provides variable currents instead of just on-off current, and the rotor teeth assume intermediate positions between the discrete positions established by the tooth patterns [3.11].

3.8 SINGLE-PHASE MOTORS

Single-phase motors will *run* on single-phase power but will not *start* on single-phase power. The explanation is too complicated for this book. To start a motor from a single-phase power supply, it is necessary to generate a second-phase current from the single-phase source and pass it through a separate armature winding.

One way to do this is to make the second winding of higher resistance so there is less phase lag in its current. Another way is to put a capacitor in series with the second winding to shift its time phase in the reverse direction. Usually a centrifugal switch disconnects the capacitor circuit after the motor comes up to speed. A third way, used on some small motors, is to put a short-circuited turn, a *shading coil,* around a portion of the magnetic field [Fig. 6-8]. It works in somewhat the same way as a second high-resistance winding.

Electric clocks driven by utility power—"plugged into the wall"—are driven by tiny single-phase synchronous motors which turn at an exact multiple of the line frequency. (Public utilities must maintain accurate frequency in order to send each other power on tie-lines.)

3.9 SPEED CONTROL

It is easy to vary the speed of a DC motor by varying field current and armature voltage. Field current can be varied by a simple rheostat. A Ward-Leonard motor generator set or its electronic equivalent provides variable armature voltage. This works because a DC motor armature generates a back voltage *(back EMF)* that bucks the line voltage and is almost equal to it. The small difference between the line voltage and the back voltage forces current through the small armature resistance. As load increases, a small reduction in armature speed causes a large percentage change in the difference between back voltage and line voltage. This large change in voltage difference produces a corresponding large change in armature current. Since torque is proportional to armature current, a small change in speed causes a large change in torque.

Very large DC brush motors are used in steel mills, hoists, and elsewhere because they can be so very accurately controlled in both speed and torque.

It is not so easy to vary the speed of AC motors powered from utility lines without an electronic power supply. Basic to polyphase AC motors is a rotating magnetic field generated by the stator winding and rotating at synchronous speed. (At 60 Hz, this is 3600 RPM divided by the number of magnetic pole pairs.) The rotor of a synchronous motor is locked to this field and turns at exactly the same speed, regardless of load. The rotor of a squirrel cage induction motor *slips* to a slightly lower speed, the slip speed inducing the rotor current, so the speed varies somewhat with load. The higher the rotor resistance, the more the slip. If the rotor is a winding brought out through slip rings, an adjustable resistance can be introduced and to provide for speed control. Two or more discrete speeds can be achieved by switching the stator winding portions to produce different numbers of pole pairs and thus different synchronous speeds.

3.10 STARTING LARGE MOTORS

An induction motor may be started simply by connecting it to the line, but there is a transient current several times the full load current. The accelerating torque may be undesirably high. If the motor is large, the transient current may trip circuit breakers and even lower the line voltage. If the load inertia is large, the transient current may overheat the motor. To prevent these undesirable effects, the motor may be started on reduced voltage and then switched to full voltage. Three common ways to do so are as follows:

1. Use an autotransformer to provide the reduced voltage.
2. Start the motor with the windings connected in Y and then switch the connections to Delta.

3. Provide multiple windings in the motor and switch them from series to parallel.

Electronic variable frequency drives provide variable synchronous speed. (Their voltage is proportional to frequency so that in the motor, the magnetic field magnitude is held constant.)

For low-power loads, a fixed-speed motor may be coupled to the load through a controllable slip clutch. If the load torque varies with speed as in a blower, the maximum power dissipated in the slip clutch may be only a small fraction of full-load power. On the other hand, if load torque is constant, as in a hoist, maximum clutch dissipation is full-load power.

Large DC motors are started with variable resistors—*rheostats*—in series with their armatures to limit starting current during acceleration.

Motor starters are discussed further in Section 9.6.

3.11 STEPPING MOTORS

Stepping motors are similar to AC synchronous motors. There are phase windings on the stator fed from electronic circuits that energize the different phases under any set of rules the circuit designer wishes. The rotor has many teeth torqued into alignment with the stator field, usually into tooth-to-tooth alignment. The rotor can be held stationary or can be moved one step at a time at any frequency and in either direction. It can be microstepped to intermediate tooth positions by applying variable currents to the stator windings. Stepping motors can control a load's position without feedback.

3.12 SERIES MOTORS

The speed of a DC motor with the field winding connected in series with the armature winding varies inversely with the load torque; this is a series motor. It is particularly useful in accelerating high-inertia loads such as railroad trains. AC series motors drive domestic vacuum cleaners at higher-than-synchronous speed to drive their blowers.

3.13 BRAKING

A DC motor may be braked by connecting it as a generator either to a resistor (dynamic braking) or to its power line (regenerative braking). This technique reduces the heat capacity required for friction brakes and is particularly useful for *overrunning loads* such a railroad trains going downhill.

An AC motor may also be braked by imposing DC on its armature winding, but the energy is dissipated in heat within the rotor winding.

3.14 INTERMITTENT SERVICE MOTORS

The principal basis for motor dimensions is the winding cross-section area; more power requires higher current that generates more heat, so large winding cross sections must be used to minimize resistance. However, for service in which power is used in short bursts with long idle periods between bursts, heat can be absorbed in the windings and core during the short burst without overheating; then the heat is slowly dissipated during the idle period. Such intermittent-service motors can be much smaller than continuous-duty motors of the same power. This analysis was first made by Charles "Boss" Kettering and it made the automobile electric starter practical. The other major application was in cash registers.

3.15 LINEAR MOTORS

If you imagine a motor that is cut along a radial plane and then unrolled from cylindrical to flat, you imagine a linear motor. Usually the moving side is cut to a short length and the other side is as long as the desired travel. Linear motors may be induction, synchronous, or stepping. The linear bearings must carry the magnetic force between the two parts; in a rotating motor, this force is balanced.

A linear motor can drive a linear motion load without intermediate gears or screws, thus without backlash and with minimum system inertia.

The largest linear motors under development are for railroad trains, usually in combination with a magnetic levitation system. Such trains need no wheels and are expected to provide a smooth, fast ride.

A linear stepping motor has been made, and used commercially, using magnetostriction as the driving effect. A nickel bar is enclosed in a coil. At each end of the coil, a sphincter-like hydraulic clamp encloses the bar. The clamps and coil are energized and deenergized cyclically so that the bar advances or retreats in steps, suggesting the motion of an inchworm. This motor was used for many years to move the massive grinding heads of a line of centerless grinders. It had a resolution of 0.000,1 inch and had great stiffness in the driving direction [A34].

Although not a *stepping* motor, a piezoelectric crystal can be made to extend linearly through a short distance. Such crystals are used in *active optics* to continually adjust the positions of segments of large telescope mirrors, such as in the Keck telescope in Hawaii. This continuing adjustment compensates for varying atmospheric refraction. (Here we have a semantic ambiguity between *motor* and *actuator,* [Chapter 6]. It is of no importance; a motor can be considered a form of actuator.)

3.16 LOW-INERTIA MOTORS

In servo systems that must accelerate and decelerate frequently, the lower the system inertia, the lower the driving power and the greater the system stability. Often, the motor inertia is a substantial part of the total inertia, so there is great benefit in minimizing it.

A first way to reduce motor inertia is to proportion the motor so that the rotor is long, with a small diameter, rather than short, with a large diameter. There are tradeoffs among material costs, heating, and other parameters, but the low inertia prevails in the design.

A second way is to omit the rotor slots so the rotor winding is mechanically free of the rotor iron core (all slot, no tooth). Only the winding is coupled to the output shaft, leaving the rotor iron stationary. Because the rotor core does not turn, the inertia of the motor is only the inertia of the rotor winding. A cost of doing this is to increase the radial length of the air gap by the thickness of the winding, thus increasing the magnetizing current and its heating effect. (The magnetic flux in a motor does not penetrate the slots and pass through the winding; instead, it jumps sideways from tooth to tooth, thus changing the flux linkage and generating torque. Similarly in a generator.)

With a permanent magnet field, there is no heating, just a larger magnet.

The secondary cost is making the winding self-supporting, even when transmitting torque. This is usually done by bonding the turns together with adhesive. Some induction motors made in this way use a solid copper cylinder or cup as the rotor winding.

An incidental benefit is smooth torquing, with no tendency for the motor to seek discrete positions established by relationships between stator teeth and rotor teeth, *cogging.*

A third way to minimize inertia is to make a low-speed, high-torque motor built into the load, without intermediate gears. The effect of motor inertia is proportional to the square of its speed. Furthermore, the elimination of gearing eliminates the backlash, inertia, and elasticity of the gears, shafts, and structure. Such motors are usually a pair of large-diameter rings, one a stator and the other a rotor. Often, the rotor is the external ring. Total system inertia is usually much less than with a small-diameter, high-speed, geared motor.

3.17 DISC MOTORS

A variation of the non-rotating rotor core motor is the disc motor. The air gap is axial. The rotor is a non-magnetic disc carrying the rotor winding and has an air gap on each side. This motor is typically DC with a permanent magnet field and with a winding made as an etched printed circuit with integral commutator segments.

This is a good point to tell that printed circuits, in addition to providing interconnections, can be shaped to include commutator and other switching segments, capacitor plates, spiral induction coils, shielding conductors of any two dimensional form, alphanumeric and graphic information, and relay and connector contacts. Electro-plating over the basic copper can give the desired surface characteristics.

3.18 MICROMINIATURE MOTORS

Etched silicon wafers were initially thought of only as transistor and integrated circuit structures. However, they are now the technology for a variety of miniature EMDs, including transducers and motors. An incidental benefit of this technology is the ability to include transistor circuits on the wafer integral with the EMD.

3.19 WATT-HOUR METERS

The watt-hour meter in your home is driven by a small motor whose speed is accurately proportional to the power transmitted through it. Its revolutions, proportional to energy, are counted and displayed by gears and dials.

In the motor, a first coil is energized by line voltage and a second coil is connected in series with the load, so it is energized by line current. Each coil's flux induces eddy currents in an aluminum disc. Each coil's eddy current reacts with the flux of the other coil. Thus the torque is proportional to

$$\text{watts} = \text{volts} \times \text{amperes} \times \text{power factor}$$

The flux of a permanent magnet also passes through the disc at a different location from the AC driving fluxes. As the disc rotates, it is mechanically loaded by eddy current drag proportional to speed. The resulting rotation angle is proportional to throughput energy. The disc drives a revolution counter that integrates the rotation and reads in watt-hours.

Watt-hour meters usually have a glass cover through which you can see the rotating disc and the meter reader can read the counter.

The basic instrument is for single phase. Polyphase power is measured by combining a separate instrument per phase on a common shaft.

This instrument has no approximations in its operating theory, has an extremely wide dynamic range, and is extremely accurate. Electronic systems for remotely reading the counter have been developed, but so far there is no successful electronic competitor for the instrument itself.

Watt-hour meter motors are also used to operate relay contacts in some measuring relays.

3.20 ELECTRIC POWER TO ELECTRIC POWER CONVERSION

A motor-generator set receives one mode of electric power into a motor which mechanically spins a generator to produce a second mode of electric power. Commonly, the first mode is three-phase AC from a 60 Hz utility, and the second mode is either DC or AC of a different frequency, such as 400 Hz.

Motor-generator sets with DC generators are also used as amplifiers. A signal voltage is applied to the DC generator field, and its armature output voltage is approximately proportional to the signal voltage. Some rotating amplifiers have internal feedback to increase gain.

The Ward-Leonard drive system mentioned above uses an AC induction motor of approximately constant speed mechanically driving a DC generator. The generator electrically drives a DC motor that mechanically drives the load. Varying the DC generator field varies the DC voltage applied to the DC motor. The motor speed can be varied from zero to full in both directions. The Ward-Leonard drive is widely used in elevators. Electronic variable-voltage drives are electronic equivalents of this classic EMD.

A form of motor generator is a single machine with a single winding in the same rotating armature. One set of connections is to AC power via slip rings, and another set of connections provides DC power via a commutator. The DC field is stationary.

Electronic controllable rectifiers for AC to DC and inverters for DC to AC have replaced many motor-generator sets.

3.21 OTHER ROTATING MACHINES

Examples of other rotating machines are:

1. A synchronous motor running without a mechanical load acts as a controllable capacitor for power factor correction in utility systems.

2. Synchros are small, rotating electrical machines, but are used to transmit information rather than to convert power, so they are described in Chapter 7 on transducers. (Some synchros transmit power as well as information.)

3. Rotary transformers are AC machines with wound rotors which do not spin but have adjustable angular position. The voltage induced in the rotor differs in time phase from the stator voltage, as determined by the angular position.

4. An induction generator is an induction motor driven by a prime mover above synchronous speed. Power from the prime mover enters the line as electric power. An induction generator can be used only where there is a synchronous generator on the line to provide the induction generator with magnetizing current.

5. Rotating machines can be used as amplifiers. The generator in a Ward-Leonard drive amplifies the signal input to its field and feeds the amplified signal to the output motor.

6. In an orbiting satellite, a current in a coil reacts with the earth's magnetic field to torque the satellite to correct its orientation.

7. Direct electric pumping is a form of motor and load. If a fluid is electrically conductive in a pipe of insulating material, if a magnetic field crosses the pipe, and if an electric current crosses through the liquid perpendicular to the pipe and perpendicular to the magnetic field, the liquid is a current-carrying conductor in a magnetic field and is urged along the pipe.

3.22 MOUNTING AND COUPLING

A motor (or generator) may be mounted on four feet, bolted down, and connected to its load (or driver) by a shaft coupling, or gears, or sheaves and belt or chain. This mounting permits position adjustment in several degrees of freedom.

A second form of mounting is by a flange integral with an end bell 2 in Fig. 3-1. The flange is bolted or clamped to the driven machine. Usually a coaxial cylindrical boss on the flange fits closely into a matching hole on the driven machine. Either no adjustment is possible or, in a *servo mount* for synchros, the motor can be adjusted in rotation and then clamped in position.

The matching hole may be an integral part of the driven machine, for a motor, or the driving engine for a generator. Many pumps and blowers are an integrated design of motor and load. The matching hole may be integral with the body of a gearbox. Typically, for gearboxes, a pinion is

machined into the end of the motor shaft and the assembly is manufactured and sold as a unit, a *gearmotor.*

A third form of mounting is built-in. The motor has no frame of its
own, but the stator and the rotor are separately mounted on coaxial portions of the driven machine. A special case is a large-diameter, slow-speed
motor that drives a servo load, such as an antenna gimbal, without gears.
Sometimes the assembly is inverted; that is, the rotor is external and the
stator internal. A gyroscope is driven by a stationary stator inside a spinning rotor; the motor rotor inertia is part of the gyroscope rotor inertia.

3.23 ENVIRONMENTAL PROTECTION

Motors are used in every imaginable environment, from coal mines to deep
space, and many environments are hostile. Below is a partial list of environments and the provisions made to prevent damage to the motor by
the environment and to the environment by the motor. NEMA specifies
most of these environments and the defenses to be made [G4].

3.23.1 Heat

Many motors have a built-in fan to circulate air between inside and
outside. Motors that sometimes operate at very low speed may have auxiliary cooling motors and blowers. Totally enclosed motors may have heat-
transfer ribs on their outer surface and may also have external fans on
their shafts to blow air over those ribs. The motor winding itself may have
heat-resisting insulation. The insulation and the bearing lubricant are
the only parts of a motor that are temperature sensitive.

Large utility generators may have liquid cooling, with liquids circulating through ducts in the rotor and stator, or may be enclosed in a high
thermal conductivity gas such as hydrogen.

Motors that operate only intermittently, such as car starters or cash
register motors, may use their own thermal mass to absorb the heat of an
operating cycle and then slowly dissipate it by conduction and convection.

3.23.2 Dust, Water, and Chemicals

Some motors are partially or totally enclosed to resist dust, water,
and chemicals. They may be enclosed in stainless steel to resist corrosion.
A variety of shaft seals are made to block fluid passage along a rotating
shaft.

Some motors are isolated from their chemical environments by using
a permanent magnet coupling, as in Fig. 3-4. The motor and driving permanent magnet 1 are sealed in enclosure 2. Thin, non-magnetic shell 3 is
part of the enclosure. Driven permanent magnet 4 is carried by shaft 5

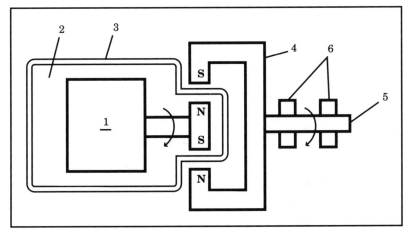

Figure 3-4   Permanent Magnet Coupling

and bearings 6. Magnetic attraction couples the two magnets. Some chemical pumps are made as a single assembly in which the pump impeller is mounted on shaft 5. Similar magnetic couplings are used to transmit motion into a vacuum chamber.

We have assumed that we must protect the motor from the environment. In some cases, we must protect the environment from contamination from the motor sparks, insulation, and lubricant.

3.23.3 Explosive Gases and Vapors

For safety we assume that the explosive gas or vapor will gradually penetrate the enclosure and be detonated by a spark from a winding defect. It is required that the enclosure prevent the internal explosion from igniting the surrounding gas or vapor.

The typical solution is, first, to make the enclosure strong enough not to burst under the internal pressure, and second, to make the leakage path for the hot gas long and thin enough that the gas is cooled to below ignition temperature by the time it reaches the outside. Fig. 3-5 shows a split cast-iron housing 1 enclosing motor 2. Flanges 3 are surface ground so that the path for hot gases escaping is very thin and puts the gases in good thermal contact with the housing. When the gases exit, they are too cool to ignite the surrounding explosive atmosphere. A similar small gap separates shaft 4 and clearance hole 5.

Such an explosion-proof enclosure is tested by filling it with the explosive vapor, immersing the assembly in the explosive vapor, and igniting the interior with a spark plug.

A second solution is to continuously flush the interior with a non-explosive gas, such as air, at a pressure slightly above ambient.

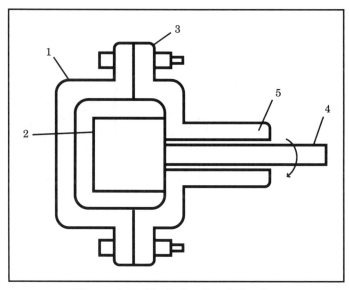

Figure 3-5   Explosion-Proof Housing

Several classes of explosive danger, together with the required safe-guards, have been defined by NEMA [G4].

3.23.4 Vacuum

The common vacuum condition for motor operation is in satellites in space. The first effect to guard against is heat; there is no cooling air, so conduction and radiation cooling must be provided. The second effect is evaporation of insulation and lubricant because even very small amounts of their vapors may condense onto optical surfaces and solar cells and degrade them.

A motor in a cooling atmosphere may drive a load in a vacuum through a magnetic coupling, Fig. 3-4, or through reciprocating bellows. It does not matter if the motor is in a small container of air and the load is in empty space or if the motor is in the earth's atmosphere and the vacuum is in a sealed chamber.

3.23.5 "Canned" Motors

The stator armature of an induction motor may be hermetically sealed by a thin, non-magnetic sleeve over its internal cylindrical surface, welded to the stator enclosure. Wires connect via hermetically sealed ter-minals penetrating the enclosure. The rotor winding, being only a low-voltage squirrel cage, need not be protected from electrical leakage and may be immersed in liquid. Such motors are used in pumps in nuclear power plants.

3.24 OTHER DRIVES

There are pneumatic and hydraulic rotary motors and rotary and linear cylinders controlled by electrically operated valves. The controlled motion may be binary, moving from one extreme to another, or it may be proportional or servo, moving smoothly anywhere between its limits. In many system designs, the best choice between electrically controlled electric power and electrically controlled fluid power is not at all obvious. Furthermore, electrical control itself competes with various forms of pneumatic, hydraulic, and mechanical control.

CHAPTER 4

Brakes and Clutches

4.1 BRAKING

Braking is either the dissipation of mechanical kinetic energy or the locking in of mechanical potential energy. Stored energy may be kinetic, as in a flywheel, or potential, as in a loaded hoist, or both, as in a loaded truck going downhill.

4.1.1 Dynamic and Regenerative Braking

As described in Chapter 3, a load may be braked by an electric generator feeding a resistive load that dissipates the energy (dynamic braking). It may also be braked by an electric generator feeding its output into a power system for use elsewhere (regenerative braking). Sometimes the stored energy originally came from a motor that acts as a generator during the braking portion of the work cycle. A braking generator may be an electric generator or a hydraulic pump. DC brush machines are usually used.

4.1.2 Mechanical Braking

Energy may also be dissipated as heat directly in a mechanical brake. This may be much more economical than by transmitting it through a generator. Furthermore, a mechanical brake may be less subject to dangerous failures than a dynamic or regenerative generator. For locking stored potential energy, such as for holding elevators in position, mechanical brakes that need not dissipate are used.

4.2 CLUTCHING

Brakes are half of what this chapter is about. The other half, clutches, are very similar to brakes in the way they work.

Mechanical power can be connected and disconnected to a load by passing it through an electrical generator, electrical controls, and an electric motor. In an electric utility feeding motors that drive machinery, in a diesel-electric locomotive, and in a submarine, the generator-to-motor system is used. However, in controlling a machine driven by a motor or engine, it is usually more compact and economical to use a clutch; a car is a common example.

Many brakes and clutches are electro-magnetically controlled, and some use electro-magnetic phenomena in their action.

4.3 POWER AND HEAT

4.3.1 Slip Power

In clutches, energy is transmitted from a source through the clutch to a load; in brakes, energy is directly converted to heat. To minimize shock, most brakes (other than locking brakes) and clutches (other than jaw clutches) are designed for a gradual transition from 100% *slip,* that is, complete disengagement, to 0% slip, that is, complete engagement. At 0% slip and at 100% slip, no power is dissipated; at intermediate slips, some power is dissipated as heat in the clutch or brake. It is its capability to *dissipate slip power* without destructive temperature rise that is the principal basis for rating a clutch or brake.

Slip power, which must be dissipated, is proportional to slip speed times torque. Slip energy is the time integral of slip power. Since torque may not be the same at all slip speeds, computing this slip energy may require graphical, numerical, or analytical integration. The brake or clutch must absorb or dissipate this slip energy without destructive temperature rise. The law of the conservation of energy applies: Slip energy must be dissipated or stored; it cannot be destroyed.

Some clutches are placed between a constant speed motor and its load and used with continual slip as speed controllers; some brakes are used with continuous slip as speed controllers with loads that are energy sources. In both cases, slip power continually generates heat within the clutch or brake, and the clutch or brake must dissipate that power without overheating.

Duty cycle is a necessary part of heating calculations, since heat is generated intermittently during slip periods and is dissipated continually at a rate dependent on temperature rise.

4.3.2 Cooling

It is necessary to remove slip heat from clutches and brakes. The following cooling techniques are used:

1. Conduction, convection, and radiation from the clutch or brake to the mounting surface, to ambient air, and to surrounding space. Fins may be provided to increase the surface area in contact with the air.

2. Fan cooling. Rotating portions of brakes and clutches often have radial fins that centrifuge ambient air over their surfaces. Many brake discs are divided into three layers with the center layer being centrifugal blower vanes joining the two friction layers.

3. Liquid cooling. Some continuous operation devices use water or oil cooling. Among these are prony brakes *(absorption dynamometers)* used as test loads on motors or engines and truck retarders that dissipate the energy of heavy trucks on long downhill runs.

a) Water is circulated through cooling passages and through a radiator.

b) The oil of oil-immersed clutches and brakes is circulated through an external radiator.

c) A water chamber is adjacent to a friction surface, and the water boils away.

Safety brakes, as in a passenger elevator or a hoist, are not used to dissipate energy, except in emergency, but switch from full off to full on at zero speed. Similarly, jaw clutches switch from full on to full off at zero speed, also without heat dissipation.

4.4 TORQUE GENERATION

Torque is transmitted to ground in brakes and to loads in clutches by the following means:

1. Friction (dry or lubricated):
Single disc pair or multiple disc
Band
Spring wrap
2. Positive:
Jaws
Shear pin
Cam
3. Viscous drag
4. Turbulence
5. Liquid inertia

6. Magnetic particle

7. Eddy current

8. Hysteresis

4.4.1 Friction

We all live with friction so it would be pedantic to write a formal definition. Reciting a few facts will be useful, however.

Friction force is proportional to normal force, regardless of area. This seems plausible if we consider that if we double the area and keep the same normal force, the force per unit area is halved but the number of units of area is doubled. This is true of both starting friction—*stiction*—and sliding friction. Starting friction is greater than sliding friction, other things being equal. *Coefficient of friction* is the name of the proportionality factor [Section 2.7].

Although mechanical friction is the dual of electrical resistance, dry friction force is non-uniform and non-linear. Surfaces are subject to wear, with which friction also varies, and the coefficient of friction varies with temperature. Friction is not accurately predictable. Lubricated friction has less wear but has all the other non-predictabilities. Electrical resistance, on the other hand, varies only with temperature, and that variation is accurately known and constant, so its effect on current and voltage is accurately predictable. (This difference is a major reason why electrical engineering can be more analytical and less pragmatic than some mechanical engineering.)

Figure 4-1 shows a single pair of friction discs coupling shaft 1 to shaft 2 in a clutch. In a brake, one of the shafts is fixed. Disc 3 is fixed to shaft 2. Disc 4 is axially slidable on shaft 1 but is angularly fixed to shaft 1 by spline 5. Control force F either presses the discs together or draws them apart.

Friction surfaces 6, the *linings,* are chosen for their friction and wear properties rather than for their strength. Linings are made of many materials including steel, cast iron, copper alloys, organic materials, and sintered metal powders. Asbestos was widely used before it was discovered to be carcinogenic. The lining is sometimes cut into segments to minimize warping from heat.

Please note that, in these sketches, a host of details required in real devices has been omitted, leaving only the means of torque transmission.

We have assumed that clutches and brakes deal with rotary motion, and most do. However, the friction clutches and brakes on the famous San Francisco cable cars are linear; the clutches grip steel cables moving linearly beneath the pavement, and the brakes grip the tracks. Operation is entirely manual and quite theatrical. Linear electro-magnetic brakes which grip the tracks of streetcars are mentioned elsewhere.

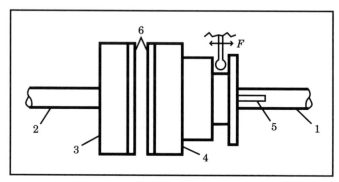

Figure 4-1   Friction Disc Clutch or Brake

Among the details of construction of brakes and clutches, not mentioned above, are:

1. The ubiquitous F that applies force between friction surfaces may be magnetic, electro-magnetic, hydraulic, pneumatic, centrifugal, or mechanical. If hydraulic or pneumatic, the oil or air may be controlled by one or another kind of EM valve, such as described in later chapters. Typically, a spring opposes an actuator so that a single command to the actuator can engage or disengage.

2. Bearings and their support, alignment, lubrication, and guarding means

3. Fluid seals

4. Mechanical structure, materials, parting surfaces, and fasteners

5. Electrical insulation of coils and their leads

6. Magnetic materials and magnetic circuit gaps

7. Bonding of friction material to support surface

8. Cooling details: heat conduction paths, cooling fins, liquid cooling circuit details

9. Coupling to input and output shafts

10. Mounting

11. All the other considerations of Part 1 of this book

The shapes of friction surfaces may be a pair of discs as shown in Fig. 4-1, a pair of matching portions of cones, a drum with a movable brake shoe as in the railroad drum brake of Fig. 4-2, a flexible band as in the band brake of Fig. 4-3, or a disc brake as in Fig. 4-4.

An automobile drum brake and a bicycle coaster brake have the same configuration as the drum brake of Fig. 4-2 except that the brake shoes

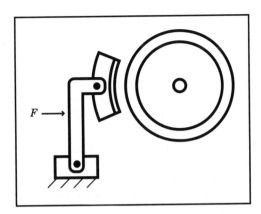

Figure 4-2 Drum Brake

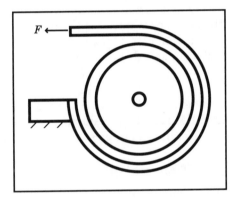

Figure 4-3 Band Brake

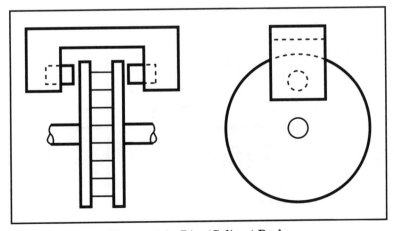

Figure 4-4 Disc (Caliper) Brake

are inside the drum. A hand-operated bicycle rim brake has the same configuration as the disc brake of Fig. 4-4.

Multiple disc devices have alternate driving and driven discs, one set splined to a shaft and the other set splined to a housing, as in Fig. 4-5. Multiple discs spread the friction torque over a large area without requiring a large diameter. They are usually immersed in oil. Lubricated friction takes place in a brake or clutch immersed in oil. The oil reduces wear and transfers heat.

Some brakes and clutches use an annular air bag to press friction shoes against a smooth cylinder.

Spring brakes and clutches use the friction between a helical spring and a smooth drum around which it is wrapped. The friction tightens the spring around the drum, so the grip is self-energizing and grabs as soon as it is initiated. The grip is released when the leading edge of the spring is blocked by an *interposer* and the self-tightening is disengaged. Retracting the interposer initiates the grip.

The action of a spring clutch is similar to the gripping of a rope coiled around a smooth drum; a light pull on the leading end of the rope is amplified to a large pull on the trailing part. (This is one of many kinds of all-mechanical amplifier [Chapter 5]. Although we are accustomed to thinking of amplifiers as electronic, there are many kinds of all-mechanical and EM amplifiers. All clutches are amplifiers; a small control power modulates a large power source to produce a large output power.)

4.4.2 Magnetic Particle

A magnetic particle clutch or brake, Fig. 4-6, has iron particles 1 suspended in oil between a pair of iron discs 2. Sometimes graphite particles are used instead of oil. An adjustable magnetic field 3 produced by coil 4

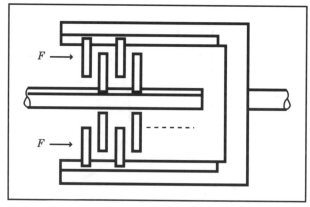

Figure 4-5 Multiple-Disc Clutch or Brake

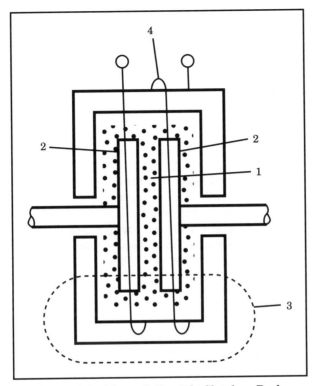

Figure 4-6   Magnetic Particle Clutch or Brake

crosses the gap between discs 2. The magnetic field causes the iron particles 1 to adhere to each other and to the discs, providing an adjustable lubricated friction device with no moving control parts. The degree of adhesion is a function of the magnetic field intensity.

An analogous device uses electro-rheological liquid and an electrostatic field that controls the viscosity of the liquid (that is what *electrorheological* means). It is still largely a laboratory device because of the low torques produced and the high voltages needed. (Historically, the electro-static device came first and inspired Jacob Rabinow to invent the electro-magnetic device as its electro-magnetic dual.)

4.4.3 Viscous Drag

Viscous drag devices transmit torque by shearing oil between closely spaced discs. The spacing may be varied or the wetted area may be varied to change the amount of drag. The oil may be pumped to transfer heat. The devices can be made in single- or multiple-disc configurations similar to Figs. 4-1 and 4-5. The discs never touch, so there is no wear.

4.4.4 Liquid Inertia

Liquid inertia devices *(fluid drives)* are related to viscous drag devices. They transmit torque by the driving rotor imparting kinetic energy to the oil and the oil imparting the energy to the driven rotor as a kind of hydraulic generator-motor. Solid moving parts never touch, so there is no wear. Slip power heats the liquid. Some fluid drives have more than one driven rotor and act as torque converters.

4.4.5 Eddy Current

Figure 4-7 shows an eddy current brake. A copper or aluminum disc 1 rotates between the poles 2a,2b of an adjustable-current electro-magnet. The disc acts as a homopolar generator that is short-circuited within itself, as in the braking portion of a watt-hour meter disc. Electric currents induced within the disk dissipate slip power as heat. Torque is proportional to speed times field strength. There is no wear.

4.4.6 Magnetic Slip Rings

Figure 4-8 illustrates magnetic slip rings. Shaft 1 is to be clutched to shaft 2 by friction between disc 3 and faces 4 on rotor 5. Disc 3 is magnetic and is free to slide axially against faces 4 on magnetic pole rings 7a,7b spaced by non-magnetic ring 8.

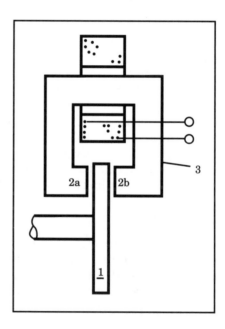

Figure 4-7   Eddy Current Brake

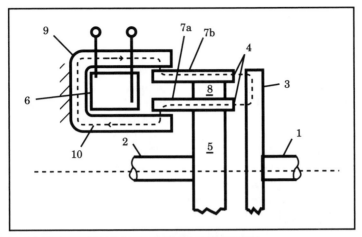

Figure 4-8   Magnetic Slip Rings

Instead of winding coil 6 on rotor 5 and conducting current into it by slip rings subject to wear and failure, the coil is in stationary core 9 whose ends overlap rings 7a,7b. Magnetic circuit 10 includes the narrow gaps between magnetic core 9 and magnetic rings 7a,7b. Magnetic rings 7a,7b and overlapping ends of core 9 constitute *magnetic slip rings*. An application of magnetic slip rings is in converting eddy current brake Fig. 4-7 to an eddy current clutch.

Rotating magnetic slip rings do not dissipate any power because no portion encounters time-varying flux linkage. One line of flux is as good as another. Conversely, in Fig. 4-7, each portion of the disc encounters a cycle of varying flux linkage each revolution, and so currents are induced and power is dissipated.

Figure 4-8 shows a single DC coil on one side of the rotation. Please imagine AC instead of DC and a coil on each side of the rotation. This is the rotary transformer that replaced brushes and slip rings in synchros.

4.4.7 Hysteresis

Hysteresis devices are similar to eddy current devices but have rotors of high-hysteresis material instead of low-resistivity material. They rotate between the poles of an adjustable-current electro-magnet. Torque is approximately proportional to field strength and is independent of speed. There is no wear. If torque is below a certain level, the two parts will lock together and there will be no slip. Under this condition, the hysteresis clutch becomes an overload slip clutch.

In both hysteresis and eddy current devices, the lossy disc and axial flux can be replaced by a lossy cup and radial flux.

4.4.8 Positive Engagement

Jaw clutches provide positive engagement, incapable of slip without breaking their teeth. They are engaged and disengaged only at zero slip and are non-dissipative. They are simple, small, and do not wear.

Overload clutches are positive engagement clutches in that the torque is transmitted via a cam retained by a spring. If torque exceeds the spring setting, the clutch instantly disengages. All-mechanical torque-limiting clutches are analogous to electrical circuit breakers.

A shear pin, shown in Fig. 4-9, is the analog of an electrical fuse. Physically, it is a cylinder of soft metal 1 that couples a hub 2 to a shaft 3, so it is a safety clutch. If torque exceeds the shear strength of the pin, the pin shears into three parts and releases the coupling. Reengaging requires replacement of the pin.

4.5 ACTUATION

Both brakes and clutches can be engaged or disengaged electro-mechanically, electro-magnetically, mechanically, or by fluid pressure. The fluid pressure is often controlled electro-mechanically by a valve.

4.5.1 Spring Actuation

Often the engaging or disengaging means for a brake operates against a spring as a restoring force so that the brake is fail-safe; a spring failure is much less probable than a failure in a controlled source of power. When you are in an elevator at the 50th floor and power goes out, it is better to be stranded at the 50th floor than to fall to the basement. Some clutches are spring engaged or disengaged, depending on the particular machine.

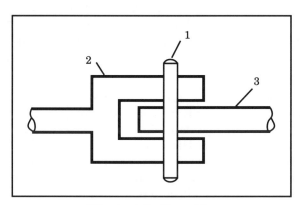

Figure 4-9 Shear Pin

Some devices have been made in which the safe condition is maintained by a permanent magnet instead of a spring. Actuation is by a coil that bucks down the magnet field.

4.5.2 Mechanical Actuation

The clutch operated by a foot pedal in a stick-shift car is a mechanical displacement clutch engaged by a spring and disengaged by foot pressure. The *emergency brake* is mechanically operated by a lever and cables.

4.5.3 Solenoid Actuation

Safety brakes, such as on that elevator, are typically dry-friction drum brakes as in Fig. 4-2 set by a spring and released by a solenoid.

Vehicle wheel brakes are dry-friction disc or drum brakes powered by hydraulic pressure, air pressure, or mechanical force from foot pedal or hand lever. Most railroad brakes use dry friction between the wheels and cast-iron brake shoes; some use disc brakes on the wheel axles, and some streetcars use electro-magnets that pull themselves against the tracks, thus reducing wheel wear.

4.5.4 Centrifugal Force Actuation

Clutches engaged by centrifugal force automatically couple motors to their loads as the motors come up to speed. The motors start under no load.

The obsolescent rotary telephone dial has a speed controller comprising a centrifugally actuated drum brake, resisting a drive spring.

4.5.5 Fluid Pressure Actuation

Pneumatic or hydraulic pressure displaces the friction discs or shoes of large friction brakes and clutches. The fluid pressure is often controlled by EM valves. The hydraulic brakes in a car use hydraulic fluid to transmit the force of the foot pedal, as amplified by its power booster.

4.5.6 Electro-Magnet Actuation

A common form of on-off friction brake or clutch has the moving disc pulled by an electro-magnet built into the device, as in Fig. 4-8. In some, the engaged state is caused by a spring, and the electro-magnet disengages the spring.

It is common to combine a clutch and a brake in a single package so that a load can be switched between a drive motor and a brake.

4.5.7 Magnetic Field Actuation

Magnetic particle, eddy current, and hysteresis clutches and brakes are controlled by the strength of their magnetic fields and have no moving control parts.

4.6 OTHER CLUTCHES

Slip clutches, as in Fig. 4-1, are safety devices with adjustable spring pressure. They slip when load torque exceeds a set point. They overheat if the drive continues to run and those using friction are inaccurate in setting, but are cheap. Hysteresis slip clutches can be set more accurately and do not wear.

A unidirectional clutch or brake, Fig. 4-10, uses a ring of rollers on spiral tracks, or of eccentric rollers between coaxial steel cylinders 3,4. In one relative direction, the cylinders turn freely with respect to each other; in the other relative direction, the rollers or eccentric rollers lock the cylinders together. The device is a mechanical diode.

The simplest—and most reliable—torque limiting clutch is a shear pin coupling a hub to a shaft, Fig. 4-9, as described above. Overload shears the pin and coupling disappears. The electrical analog is a fuse, but fuses can have thermal lag that permits momentary overloads; the release of a shear pin is instantaneous.

There are numbers of other clutches and brakes. Some are automatic and triggered by torque or speed or direction and some are operated electro-mechanically.

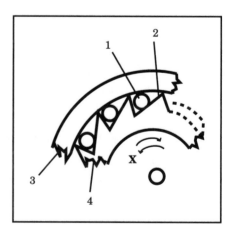

Figure 4-10 Unidirectional Clutch

CHAPTER 5

Amplifiers

5.1 GENERAL

5.1.1 Amplifiers

Most of us associate the word *amplifier* with electronics, but there are many other kinds of amplifiers. An amplifier is any device that has a *power source,* receives a *control signal* that modulates source power, and emits the result as *output power.* The amplifier *amplifies,* or increases, the control signal power into output power. The amplifier draws power from the power source, since power cannot be created or destroyed. The power source may be called the *power input,* and the control signal may be called the *signal input* or *control input.*

Output power is a function of control signal, but not necessarily proportional to it, as the examples below illustrate. Several control inputs may be combined by an *operational amplifier.* Although the output power is usually greater than the control signal power, it may be at any impedance level; for example, an electronic amplifier output power voltage may be less than its control signal voltage but may deliver much larger current.

In the same amplifier, the signal input, power source, and output power may each have a different form of energy: electrical, mechanical, pneumatic, hydraulic, or thermal.

Examples of amplifiers are the following:

1. An automobile power brake is an amplifier: Its *power source* is a hydraulic pump, its *control signal* is mechanical force from the brake pedal, and its *output power* is greater force through the wheel brake cylinders. It becomes, after a brief transient, a force amplifier rather than a power amplifier: The input force on the brake pedal is amplified to the output force on the brake shoe, with no continuing energy flow.

2. An automobile power-steering motor is a similar amplifier that amplifies the power from your hands into the steering wheel into the greater output power used to turn the automobile's front wheels.

3. A light switch is an amplifier. The power source is the utility voltage, the control signal is your hand mechanically moving the switch lever, and the output power is the electric power to the light. The output power is not *proportional* to the control signal; both are binary, either on or off. Such a two-condition system is sometimes called a *bang-bang* amplifier.

4. There are systems is which the control signal varies smoothly over a range but the amplifier output power switches from fully off to fully on at one input value and from fully on to fully off at a different value.

5. In electronic *switching amplifiers,* the output power is rapidly switched on and off from the power source. The average output power varies with the average time the switch is on and is smoothed by a filter. Switching is usually at high frequency. Switching amplifiers have very low losses; power is never divided between useful output and internal dissipation. Other electronic amplifiers are dissipative, as are hydraulic amplifiers.

(Mechanical and fluid switching are slower than electrical, so there are no practical mechanical or fluid switching amplifiers.)

6. All fluid valves are amplifiers, and there is an enormous variety of valves. Some are bang-bang and some are *throttling;* that is, they vary the magnitude of output power as a function of the magnitude of control signal. Some signal inputs are human, some are mechanical motions, some are electrical, some are fluid pressure, some are temperature. The power source and output power may be liquid, gas, or vacuum. (Vacuum tubes are valves for electricity, and in England they are called valves.)

7. And, of course, there are proportional amplifiers in which the output power magnitude is simply proportional to the control signal magnitude.

We have spoken of output "power." The controlled output parameter may really be voltage or pressure or position or speed or other, rather than "power," and the actual output "power" (the rate of energy flow) may also be a function of a varying load.

More kinds of amplifiers are described below.

5.1.2 Forms of Energy

An amplifier's control signal, power source, and output power need not all have the same form of energy, that is, electric, mechanical, hydraulic, pneumatic, thermal, optical, or acoustic. For example, your TV set has amplifiers responsive to mechanical control signals in its control knobs and to electro-magnetic radiation control signals into its antenna; it has an electrical utility power source, and it puts out both acoustic and optical output power. The optical output power is the result of several cascaded amplifiers, the last of which is the picture tube. The acoustic output power is also the result of several cascaded amplifiers. (But the final acoustic element, the loudspeaker, is not an amplifier at all but is an EM *actuator* [5.1.4]. It has electric input and acoustic output, but no power source modulated by the electric input.)

5.1.3 Mechanical Advantage Device

A *mechanical advantage device* receives mechanical force and displacement and emits mechanical force and displacement, but it has no separate source of power; it is not an amplifier. For example, an automobile hand-brake lever receives mechanical force at the handle and emits a larger force to the brake cables. It has no other source of power. Mechanical advantage devices include levers, gears, and pulleys, the "elementary machines" of a physics book.

Electrical transformers are electrical duals of mechanical gears. Air and oil intensifiers are air and oil duals of mechanical levers.

A transformer is an impedance changer in that it changes the ratio of volts to amperes without changing the product. Similarly, mechanical advantage devices and fluid intensifiers may be considered mechanical and fluid impedance changers.

5.1.4 Actuators

An *actuator* receives energy in one form, emits the same energy in another form (minus losses), but has no separate power source; for example, the loudspeaker mentioned in [5.1.2]. Brake cylinders are actuators with hydraulic inputs and mechanical outputs. A garage-door opener is an actuator with electrical input power to its motor and mechanical output power to the garage door.

Some actuators are packaged with input amplifiers. For example, a typical garage-door opener assembly includes an amplifier comprising control circuitry and a connection to utility power.

The word *actuator* is usually applied to devices whose outputs are intermittent rather than continuous, for example a door operator rather than a motor driving a pump.

Chapter 6 deals with actuators at greater length.

5.2 CLUTCHES

Clutches [Chapter 4] are amplifiers. Control signals used include mechanical displacement, hydraulic pressure, pneumatic pressure, and electric current. Torque is transmitted via dry friction, lubricated friction, viscous shear, and electro-magnetic force. The power source is mechanical shaft rotation, and the output power is mechanical shaft rotation. For example, the clutch in a stick-shift car is an amplifier that has foot position as the control signal, crankshaft rotation as the source power, and driveshaft rotation as output power.

Clutches are dissipative amplifiers, unlike the electrical switching amplifiers described above. Input power is proportional to input speed times torque. Output power is proportional to output speed times the same torque. The difference in speed is called slip. When there is slip, output power is less than input power and the difference, slip speed times torque (slip power), appears as heat. Clutches are valuable for producing transient accelerations, as in a car, but they make poor speed controllers because they get hot from dissipated slip power.

There are small clutches that use impact on an interposer (see below) instead of friction to absorb energy. The input is a mechanical trigger. They are used in single-revolution drives.

Electronic amplifiers with very rapid switching between on and off, and with *pulse-width modulation,* provide variable-speed mechanical power when they feed electric motors. Electrical inductance and mechanical inertia smooth the power flow. The benefit of such switching amplifiers is that they are non-dissipative, which more than makes up for their complexity.

5.3 VARIABLE-RATIO DRIVES

There is a class of all-mechanical amplifiers in which the control input varies the ratio of input power displacement and speed to output power displacement and speed. (This is analogous to varying electrical impedance ratio.) An example is a V-belt and pulley drive having a variable pulley diameter. Another example is a manual shift gearbox (analogous to a transformer with taps on its windings). An automatic automobile transmission is a variable-ratio gearbox with slipping brakes to generate a smooth transition from ratio to ratio.

The automobile automatic transmission is an amplifier with an automatic control input. The manual shift gearbox is an amplifier with a human control input.

5.4 BRAKES

Brakes are the same as clutches except that the output is locked, so output speed is always zero. It is the slip power dissipation described above which makes brakes useful for gradually slowing or stopping inertia loads, but overloaded brakes get hot from the same slip speed which makes clutches get hot! There are electrical and hydraulic braking systems that convert slip power to electrical or hydraulic power and conduct it away for dissipation elsewhere. A water-filled prony brake in a laboratory converts slip power to boiled water and dissipates it as steam.

5.5 VALVES

Valves are amplifiers that have fluid pressure and flow as power source, fluid pressure and flow as power output, and either mechanical displacement, fluid pressure, or electro-magnetic force as control signal. Non-mechanical control signals are first converted, inside the valve, to mechanical displacements which throttle the flow conduit. There are many, many kinds of valves.

A particularly useful valve amplifier is shown in Fig. 5-1. The figure is divided into three portions. Portion JP is a *jet pipe valve* swung left and right by a polarized electro-magnet such as shown in Fig. 6-4 and described in [Section 6.3.3].

Portion SV is the *servovalve* we shall discuss here.

Portion PC is the *power cylinder,* which is the output actuator driven by the servovalve.

Command information enters the polarized electro-magnet as a low-power electric signal. The electro-magnet rotates the jet pipe clockwise or counterclockwise, depending on the polarity of the electrical signal. The jet pipe drives the servovalve left or right with impact pressure of the oil jet, and the servovalve drives the power cylinder left or right, as explained below.

Assume a hydraulic power supply that provides oil under pressure, P, and drains back oil at atmospheric pressure, D. Servovalve SV comprises a cylindrical cavity 1 in which slides spool 2. Spool 2 has three flanges 3,4,5 that fit closely in cavity 1. Ports 6a,6b connect to the jet pipe assembly JP so that pressure on port 6a drives the spool to the right and pressure on port 6b drives the spool to the left.

When the spool is centered, as shown, ports 7 and 8a, 8b are blocked by the spool flanges.

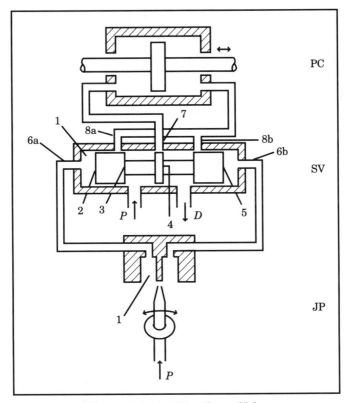

Figure 5-1   Jet Pipe Servo Valve

When the spool moves to the left, pressure from port P is connected to port 8a and thence to the right end of power cylinder PC, and the left end of PC is connected to the drain port D via port 7.

When the spool moves to the right, pressure from port P is connected to port 7 and thence to the left end of PC and the right end of PC is connected to the drain port D via port 8b. PC moves left and right with a large force. (Figure 5-1 is not to scale; typically, SV is a fraction of an inch in diameter and the spool moves a very small fraction of an inch. PC may be several inches in diameter and may move several feet.)

In another version of the hydraulic servo valve, the polarized electromagnet mechanically drives a first, tiny, servo valve that drives SV as a second-stage amplifier.

5.6 ENGINES

Engines are amplifiers whose control signals are mechanical throttle positions. In carbureted internal combustion engines, throttle valves vary the flow of the fuel and air which are the power source. In diesel engines, the throttle controls the injection fuel pump stroke. In steam engines, reciprocating or turbine, throttle valves vary steam flow instead of fuel flow. In water engines (water turbines), throttle valves vary water flow.

Actually, it is the throttle that is the amplifier; the engine is an actuator that converts all its input power from the throttle to output power.

Electric motors are engines with electricity as the power source; throttling is either by an external electrical control or by the internal self-regulating behavior of the motor itself. In each case, power output is mechanical rotation.

Rockets are engines whose power source is liquid or solid fuel and oxidizer. With liquid fuel, the output power level is controlled by throttling the fuel and oxidizer. (The mechanical output is the reaction momentum into the vehicle balancing the momentum of the exhaust gas.)

5.7 FLUID VALVES

Fluid power uses compressed air, pressurized oil or water, or steam as energy sources.

Most fluid power amplifiers are valves. Most actuators are cylinders or fluid motors. Fluid power amplifier valves are made in a great variety of types and sizes. Some are on-off and some are throttling. Throttling valves for fluid power are dissipative, on-off valves are not.

There are valveless hydraulic amplifiers which are variable flow pumps. Control signals adjust the pump displacements to vary their output flows. These are non-dissipative amplifiers.

5.8 INTERPOSERS

An interposer amplifier is a mechanical stop positioned by a control signal. The interposer is put into the path of a body and prevents a force from moving the body. If the force is continually applied to the interposer through the body, the interposer is a *trigger,* as in a gun. An interposer may also be put into the path of the body before the body reaches it, and determines whether or not the body shall pass. A plurality of interposers provides a plurality of stop positions.

5.9 PYROTECHNICS

A pyrotechnic actuator uses a self-contained burning fuel like gunpowder for its power source and a control signal from either an electric current or a mechanical impact to ignite it. The electrical control signal usually heats a resistor, called a *carbon bridge,* to incandescence to ignite the charge. Usually, two redundant resistors are provided for reliability.

The output is mechanical, usually a fluid power cylinder. If the cylinder piston is a projectile, the pyrotechnic is a gun. An enormous amount of energy can be controlled by a very small control signal. Once started, all the stored energy in the power source must be expended at once.

Military mines, torpedoes, gun rounds, and missiles are pyrotechnics. Their burning rates vary from slow, in the case of rocket engines, to fast, in the case of gun propellants, to very fast, in the case of warheads.

The control amplifiers for military pyrotechnics contain *safety devices* to prevent accidental firing, *arming* devices to cancel the safety features and prepare the fuse to fire, and *fuses* to fire the charge at a designated stimulus. The arming device may integrate time or acceleration, so the fuse is not enabled until the missile is at a distance from the launcher.

Safeing, arming, and fuzing is an elaborate art using mechanical, electrical, and chemical technologies, usually in combination [Chapter 15].

A variant of pyrotechnic amplifier is the fusible link. Instead of heating a carbon bridge, the control signal melts a link of low melting point alloy such as used in electrical fuses. The link had been under tension or shear from a spring, whose stored energy is the amplifier power input. The melted link frees the spring to provide the output power.

5.10 FEEDBACK CONTROL

An amplifier's output power can be controlled to a desired level by measuring its output with a sensor whose own output reaches back and combines with the control signal.

The first *feedback control system* was made by James Watt for his steam engine. He invented the flyball governor whose speed is the engine output speed and whose output adjusts the engine throttle, Fig. 7-14. Since that day in the eighteenth century, the art of feedback control for all kinds of amplifiers has grown enormously. Many engineers specialize in the subject, which has become highly mathematical. The word *cybernetics* was coined by Norbert Wiener for his introductory book on the subject [A29]. *(Cybernetics* means feedback control; it was derived from the Greek word for steersman; some early feedback control amplifiers were used for steering ships.) Another word for a feedback control system is *servomechanism* or *servo*. There are so many texts on the subject that I have not included any others in the references.

Figure 5-2 shows a feedback control system in which electrical amplifier 1 receives an electrical control signal 2, representing a position command, from source 3 and electric power from source 4. Electrical output power 5 is the control signal to cascaded electro-hydraulic amplifier 6 that also receives hydraulic power from source 7. Hydraulic output

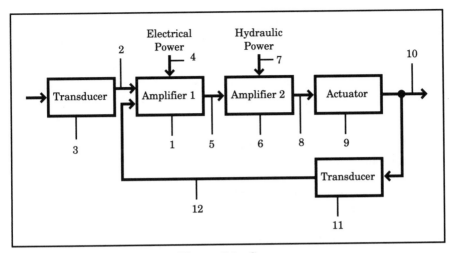

Figure 5-2   Servo

power 8 is the sole input to hydraulic actuator 9 (perhaps a cylinder and piston), whose output 10 positions a load (perhaps a ship's rudder.)

Transducer 11 measures the actual mechanical position of load 10, translates *(transduces)* it to an electrical feedback voltage 12 that is a second input to amplifier 1. The *difference* between command voltage 2 and feedback voltage 12 is amplified by amplifier 1.

5.11 CASCADED AMPLIFIERS

Amplifiers are often cascaded, one feeding another, either for more power or for control or for changing the form of energy. Examples:

Steam boilers are amplifiers cascaded ahead of steam engines. Their power source is heat from either burning fuel or fissioning atoms, their feedback sensors measure pressure, their control signals throttle either fuel and air or neutron flux, and their output power is pressurized steam.

In a steam power plant, the steam boiler output is the *power source* for the steam turbine throttle amplifier. The turbine output is the *power source* for the electric generator, whose output is the product of the power plant and is the *power source* of its customers. Electrical frequency is the feedback control to the throttle amplifier.

In an automobile power steering system, a manual signal to the steering wheel is amplified by the hydraulic steering motor whose output power moves the front wheel steering linkage. Then the front wheel angle, in turn, is the amplifier which steers the car, using the car's engine as a power source.

In an electronic analog computer, an *operational amplifier* may receive several control signals. Its output is proportional to the algebraic sum of the signals. The operational amplifier may be made to integrate or differentiate or perform other mathematical functions on the signals. An analog computer contains a network of such amplifiers made to behave as an analog of a mechanical system, with voltages and currents corresponding to forces and displacements. The usual purpose of analog computers is to simulate, not to provide amplified power, but analog computers may be used as equipment controllers and be provided with cascaded power amplifiers to control actuators.

Electronic analog computers were preceded by mechanical analog computers in which quantities were represented by shaft rotations, gear differentials added and subtracted, ball-and-disk integrators did integration, and linkages did trigonometry. Artillery fire control and bombsight computers were of this class well into World War II. Most analog computers have been superseded by digital computers.

In an electronic communication amplifier, the control signal may modulate a power source to produce a signal of a different form, such as amplitude modulated (AM) or frequency modulated (FM) radio frequency (RF) output power.

5.12 FLUIDICS

Fluidic amplifiers use a steadily flowing fluid as the power source and a time-varying cross flow of fluid, impinging on the first flow, as the control signal. The output is the switched path of the power source fluid. The fluid may be liquid or gas.

The modern auto windshield wiper is controlled by an fluidic air valve.

5.13 SWITCHES

Switches are amplifiers that transfer power from one conduit to another or that turn power on and off. Valves switch fluid flow, metal contacts or semiconductors switch electricity, and mechanical switches transfer motion in railroads and conveyors from one track to another. Switch amplifiers do not modulate output magnitude, but they change output path. Electronic *switching amplifiers* perform *pulse-width modulation* by varying the ratio of on-time to off time.

In descriptions of mechanical systems the word *amplifier* is not usually used, although the word *actuator* is common. Nevertheless, it will be useful to you in designing products in which energy flows and force is applied to recognize the relationships among amplifiers and actuators.

This chapter has extended the concept of "amplifier" beyond its common limitation to electrical devices. The intent is to provide you a deeper insight into EM device engineering.

CHAPTER 6

Actuators

6.1 ACTUATORS AND AMPLIFIERS COMPARED

Actuators differ from amplifiers. Both receive input power and emit output power, *but:*

An amplifier requires only a low-power control signal. The signal modulates a power source to generate output power. An actuator has no power source other than its control signal, which must be powerful enough to provide the output power plus losses.

Such drives as a DC shunt motor or an AC induction motor maintain approximately fixed speed and, by themselves, draw just the amount of power from constant voltage inputs to provide the torque needed by a varying load. They are not usually referred to as actuators unless they are fed with variable voltage from control amplifiers. To modulate the output power of most actuators one must modulate the output power of an amplifier that feeds the actuator.

This distinction between amplifier and actuator seems pedantic until one considers examples:

1. A radio receiver has a succession of electronic *amplifiers* until the amplified signal reaches the output *actuator,* the loudspeaker. The loudspeaker is not electronic at all but is electro-mechanical. It converts all the modulated electric power from the last audio amplifier into acoustic power, minus losses.

2. A positioning servomechanism has a transducer, an amplifier, and a motor. Transducers will be described in the next chapter, but they generally have a non-electronic input from the controlled variable and an electronic output that is a feedback signal to the amplifier. The actuator is a motor that converts all the modulated power from the electronic amplifier into mechanical power to the positioned machine.

3. A steam power plant has a speed transducer *(governor),* Fig. 7-14, whose feedback signal goes to the throttle valve, which is an *amplifier,* as we saw in Chapter 5. (Input power: pressurized steam from a boiler, control signals: desired speed-setting displacement and governor output displacement, output power: pressurized steam.) The output power goes to a spinning turbine, an *actuator,* that converts all the steam power from the amplifier (throttle valve) into mechanical power. A second *actuator,* cascaded to the output of the first, is the alternator that converts all the mechanical output power of the turbine to its own electrical output power.

6.2 EM ACTUATOR PHENOMENA

EM actuators use the following phenomena:

1. Electro-magnetic force
2. Electro-static force
3. Piezoelectric force
4. Magnetostrictive force

Electro-magnetic force is used in motors with unlimited rotation [Chapter 3] and in several kinds of limited displacement electro-magnets described below. Most EM actuators are electro-magnetic.

Actuators using the other three phenomena are described later in this chapter.

6.3 POLARIZED ELECTRO-MAGNET ACTUATORS

In these actuators, flux from a permanent magnet reacts with current in a coil. The output force or torque is proportional to the product of constant flux from the permanent magnet times the *first power* of the current. A DC shunt motor with a permanent magnet field is such an actuator.

6.3.1 Linear Voice Coil

The name comes from the coil that drives the diaphragm of a loudspeaker along a straight line. See Fig. 6-1.

A permanent magnet 1 and soft iron pole pieces 2,3 provide a fixed magnetic field within which a coil 4 moves axially. Force F is proportional to coil current, which may be DC or AC. There is no moving iron, so the only inertia is that of the coil and its mechanical load 5. The reluctance of the coil's magnetic circuit is quite large, so there is little hysteresis or eddy current loss. The force is accurately proportional to current over a wide range of magnitude and over a frequency range from DC to at least 20 kilohertz (kHz). Power capacity is limited only by coil heating.

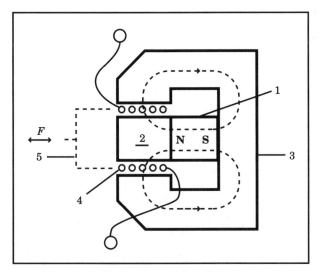

Figure 6-1   Voice Coil

Voice coils are used, for example, in

- Earphones
- Loudspeakers
- Computer disk drive head movers
- Hydraulic servo valves
- Vibration testing machines (generating up to several thousand pounds of force)

It is theoretically possible, though rarely done, to reduce the inductance of the voice coil almost to zero. A compensating winding is added, analogous to the compensating winding sometimes embedded in the pole faces of a DC motor to cancel armature reaction. The compensating winding is a stationary coaxial coil connected in series with the voice coil and opposed to it in magnetizing direction. The cost of the compensating coil is doubling the copper loss in the voice coil and in almost doubling the strength of the permanent magnet to maintain the same flux density in the wider air gap.

It is possible to blow cooling air axially through the voice coil space to increase its power capacity, which is limited only by heating.

6.3.2 d'Arsonval Galvanometer

Rotary motion "voice coils" are called *d'Arsonval galvanometers,* named after their inventor. Most operate on DC, have pointers moving across scales, and are the common electric meters such as appear on auto

114 CHAPTER 6

dashboards. (More precisely, these are *indicators* of a present quantity; *meters* integrate a quantity over time. Examples are watt-hour meters, gas meters, and odometers.)

Other d'Arsonval galvanometers include:

- Strip-chart pen recorder, Fig. 6-2
- Strip-chart mirror galvanometer recorder, Fig. 6-3
- Reflected scale galvanometer

In Fig. 6-2, coil 1 is pivoted on stationary iron core member 2 between the poles 3 of permanent magnet 4. Electrical connections 5a,5b are spiral springs similar to the balance wheel spring of a mechanical watch. These springs provide both a restoring force and an electrical connecting path. Low inertia arm 6 carries recording stylus 7 (ink or heat). Stylus 7 marks strip chart 8 as the chart is drawn from storage roll 9 by sprocket 10.

Figure 6-3 shows an analogous structure with a frequency response in the audio range. Coil 1 is a single, narrow, flexible loop of fine wire to which tiny mirror 2 is cemented. The loop rotates through small angles by bending and twisting its straight portions as flexure pivots. The light beam provides a long arm with zero inertia and writes on photographic film. The loop is usually mounted in a capsule immersed in damping oil. Usually these galvanometers are placed close together between the poles of a wide, permanent magnet. They produce a simultaneous record of many variables. The cathode-ray oscilloscope is faster, of course, but does not provide as many channels unless there is electronic switching. An-

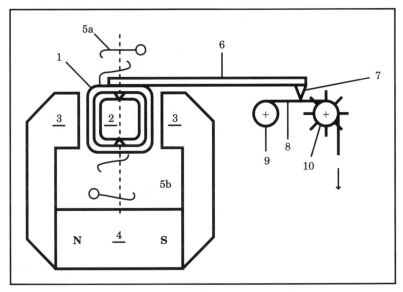

Figure 6-2   Strip Chart Pen Recorder

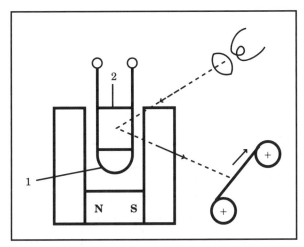

Figure 6-3 Mirror Galvanometer

other competitive technique is electronic analog-to-digital conversion and solid-state memory for each channel, followed by slow plotting. The current practice is to digitize data and record on an EM magnetic tape recorder.

A cross between the strip-chart pen recorder and the reflecting galvanometer recorder is the reflecting galvanometer. It resembles Fig. 6-2 except that arm 6 is replaced by a mirror and the pivots are replaced by a quartz thread from which the coil hangs. The mirror reflects a curved scale into a laboratory telescope. The result is a very sensitive galvanometer.

A version of the reflecting galvanometer is the ballistic galvanometer. It has a high inertia coil. If a pulse is delivered to the coil, the swing amplitude is proportional to the integral under the pulse curve. The amplitude of a full swing is observed through the telescope. One use is the measurement of a magnetic field passing through a search coil. If the search coil is rapidly removed or if the current generating the field is interrupted, the galvanometer indicates the change in flux through the search coil.

6.3.3 Iron-Core Polarized Electro-magnets

Figure 6-4 shows a *polarized electro-magnet* (that is, one with a movable iron armature, a permanent magnet, and a coil). A pivoted iron armature 1 is enclosed by a stationary coil 2 whose inside is large enough to let the armature move. The flux 4a,4b from permanent magnet 3 divides through stationary cores 5a,5b as shown, so there is equal flux in all four air gaps 6a,6b,6c,6d and there is no net torque on armature 1.

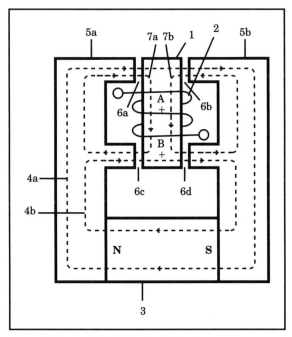

Figure 6-4 Polarized Electromagnet

Current in the coil produces a second magnetic field 7a,7b.

Fluxes 4a and 7a *add* and *increase* the force in air gap 6a.

Fluxes 4b and 7b *add* and *increase* the force in the diagonally opposite air gap 6d.

Fluxes 4b and 7a *subtract* and *reduce* the force in air gap 6c.

Fluxes 4a and 7b *subtract* and *reduce* the force in air gap 6b.

The net result is increased force in diagonally opposite gaps 6a,6d and decreased force in diagonally opposite gaps 6b,6c. These forces result in a counterclockwise torque on armature 1. A reversed current in coil 2 produces reversed fluxes 7a,7b and a clockwise torque.

Displacement of the armature from symmetry unbalances the forces when the current is removed. Therefore, a centering spring is necessary unless the application calls for bistable positioning, left and right of symmetry, as in a two-position polarized relay.

The pivot can be either centered at point *A* or at one end of the armature at point *B*. In the second case, the air gaps can be shortened at the pivot end.

Uses of polarized electro-magnets include the following:

1. Polarized electro-magnets were once used to drive loudspeaker diaphragms, but these are now driven by voice coils, which are more linear.

They are for DC only; if AC were applied to the coil in Fig. 6-4, the moving armature 1 would just oscillate at the AC frequency.

2. Polarized electro-magnets are used as the actuators of sensitive relays.

3. Polarized electro-magnets drive *electro-hydraulic servo valves.* These valves control oil flow to the two ends of a hydraulic cylinder as part of a position feedback control system, Fig. 5-1, Section 5.5.

Electro-hydraulic servo valves are typically two-stage amplifiers, each stage being a four-way valve, as in Fig. 5-1. A polarized electro-magnet moves the spool of the first stage. The reversing hydraulic output of the first stage moves the spool of the second stage. The reversing hydraulic output of the second stage moves the power piston.

The polarized electro-magnet may be coupled mechanically to the first-stage spool or it may swing a jet of oil through a *jet pipe* against one or another of two windows connected to the ends of the first-stage spool, as in Fig. 5-1.

(Usually a servo valve is mounted directly on its power cylinder to minimize the resilience of the oil in the connecting tubing. Oil really is compressible, and it is this compressibility that limits the frequency response of hydraulic servos.)

4. A polarized electro-magnet drives a valve that modulates the air flow in a compressed-air loudspeaker; see Chapter 12, "Sound."

6.4 TWO-COIL ACTUATOR

A two-coil actuator has no permanent magnet. Instead, it has one movable coil and one fixed coil. The force or torque is proportional to the product of the two-coil currents. For example:

1. In a wattmeter, the load current passes through one coil, and the load voltage forces a second current through the second coil. The torque is proportional to the product of the two currents.

2. A watt-hour meter has a current coil and a voltage coil. Each coil current induces a proportional secondary current in an aluminum disc. The secondary current reacts with the other current's flux to produce a torque *on the disc.* The disc is free to rotate without limit so it integrates the (in-phase) product of the two input currents.

3. Accurate voltmeters and ammeters are made in the same way as the wattmeter, with coils connected in series, carrying the same current, and working against a spring. Displacement of the pointer is proportional to the square of the current.

4. In analytic balances, an unknown mass was weighed by balancing its weight with a set of calibrated weights. A faster system is now used to balance the unknown weight with the force between two current-carrying coils, in series, and reading out the balance current with a digital voltmeter.

6.5 NONPOLARIZED ELECTRO-MAGNET ACTUATORS

The word *solenoid* originally meant a helical coil of wire that produced a magnetic field when a current flowed through the wire. However, the word has come to mean an assembly containing a single coil, a fixed core of soft iron, and a moving armature of soft iron. The movable core of an electro-magnet is called its *armature*. (The word is the same as for the winding of a motor or generator, but there is no connection.) This is a *nonpolarized electro-magnet.*

A coil, usually on a fixed iron core with a movable iron core somewhere in its field, causes a force on the movable core in the direction of increasing flux linkage. Fig. 6-5 is the most common configuration. The flux is proportional to current I and I is also proportional to itself, so the force is proportional to I^2. The flux is not exactly proportional to I because during armature motion the saturation of the iron changes and the leakage flux changes. Therefore, the force response of a nonpolarized electro-magnet is not exactly square law.

Compare this performance with the voice coil or polarized electro-magnet forces that are accurately proportional to I. In measuring devices and proportional controlling actuators, the polarized actuator is far superior because of its accurate linearity. Also in devices that must operate on low power, the output force, which is proportional to the product of the large, permanent magnet flux and the small control current, is greater.

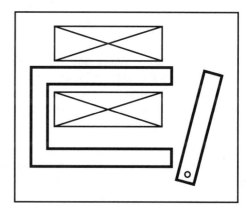

Figure 6-5 Single Coil, Pivot at One Pole

On the other hand, single-coil, square law devices are much cheaper. For on-off operation where there is adequate operating power available, they are quite suitable and are usually used.

Nonpolarized electro-magnets are used as short-stroke actuators in a myriad of uses, from relays to door latches to trigger actuators.

6.6 ELECTRO-MAGNET CONFIGURATIONS

Another technique benefiting DC electro-magnets is enlarging the ends of the poles, Fig. 6-6. Polar enlargements provide larger pulling areas without increasing core diameter and consequent coil size. When the armature closes, the core saturates. This saturation limits pull beyond what is needed, but the limitation does no harm.

A solenoid is a means to transmit force through the wall of a sealed enclosure. The magnetic flux penetrates the wall and exerts force on an armature inside. The most common use is the mercury contactor in which mercury and an iron armature are enclosed in a sealed container with the iron floating on the mercury. A pair of electrical leads penetrates the enclosure. The enclosure is surrounded by a coil, with or without an outside iron core. When the coil is energized, it pulls down the iron armature that displaces the mercury upward that closes the circuit between the penetrators.

6.7 AC AND DC ELECTRO-MAGNETS COMPARED

6.7.1 Difference in Principles

AC electro-magnets, while still square law devices, are different from DC electro-magnets, as follows:

1. The core and armature must be of laminated silicon steel to minimize hysteresis and eddy current loss. This laminated structure imposes

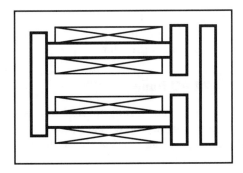

Figure 6-6 Polar Enlargements

severe limitations on the forms in which the cores can be made. (Some very small AC electro-magnets are not laminated, or have only the core but not the armature laminated, or merely have a slot in their plungers, that is, cylindrical armatures.)

2. In DC electro-magnets, the current equals applied voltage divided by coil resistance and is independent of armature position. This is not so in AC electro-magnets. Current is little affected by coil resistance. Just enough current flows to produce just enough AC flux linkage to induce just enough AC voltage in the coil to balance the applied AC voltage. (In electrical engineering language, the impedance is mostly reactive.) Therefore, coil current automatically reduces as the armature approaches the core and magnetic circuit reluctance falls. Empirically, AC electro-magnets have force/stroke characteristics between proportional and square law.

3. As a very rough approximation, because the flux field changes shape with armature position, AC electro-magnets have flux that is proportional to applied *voltage* and is independent of armature position.

This effect can be a great benefit. The AC electro-magnet draws a great surge of current when it is open, enough to burn out the coil if continued. But when the armature closes, the current automatically reduces to a safe value. The flux linkage remains constant. Thus an AC electro-magnet can pull a larger load through a longer stroke than a DC electro-magnet of the same size. All benefit is not on one side, however, as the next section shows.

A DC electro-magnet can be fitted with a limit switch near the end of its stroke to reduce steady-state current to its minimum holding value by switching a resistor into series with its coil.

Another very elegant DC technique used in a Teletype machine is to close the armature against the poles by a motor-driven cam and then either turn off the current or not turn off the current, as the control information commands. Thus the DC electro-magnet acts at its maximum force position and does not have to pull a load through a distance. This situation is analogous to the use of a DC lifting magnet, Fig. 6-7, which is lowered into contact with its load before it is raised so the magnet need never pull through a long air gap.

6.7.2 AC Hum

An AC electro-magnet hums because its flux passes through zero each half cycle, so its force also passes through zero each half cycle, although its average force carries the average load. The hum is at twice line frequency, typically 120 Hz, well within the audio range. If the load has low

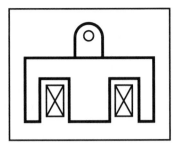

Figure 6-7 Lifting Magnet

inertia, the hum becomes a loud chatter as the armature is alternately released and attracted.

The usual solution to this problem is to provide a *shading coil,* a copper ring enclosing part of the *pole* facing the armature, Fig. 6-8. This shading coil has the same effect as the motor shading coil in Section 3.8. The flux through the shading coil induces a current in it that lags, in time phase, the flux outside the shading coil. This lagging current causes a phase shift in the flux through the shading coil. Thus when one portion of the flux is zero, the other portion is not and exerts a holding force. For minimum force reduction, the area enclosed by the shading coil is greater than the area outside it, and the area outside it is cut back slightly, as shown, to give it a greater reluctance and force more of the flux through the shading coil. This is no small matter. The permissible load for quiet operation is much smaller than the average magnetic force, and the mechanical fit of armature to pole must be parallel and close.

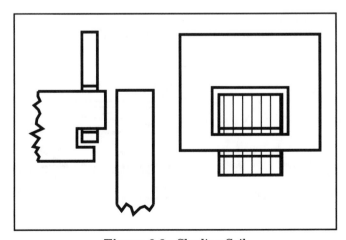

Figure 6-8 Shading Coil

6.7.3 Coils

Both DC and AC electro-magnets can be made with either a single coil or with two coils. Fig. 6-9 shows the same core wound with two coils (a) and with a single coil (b). The benefits of two-coil construction are shorter average turn length and resistance and greater surface area to dissipate heat. Therefore, there are more magnetizing ampere turns for the same temperature rise. Two coils cost more than a single coil, however. Plunger armatures solenoids, Fig. 6-10, compel a single coil.

For the same line voltage, DC electro-magnet coils have many more turns of finer wire and draw less current than AC coils. This does not indicate less power consumption, however. The AC current is mostly reactive and furthermore is automatically reduced when the armature closes, as described above.

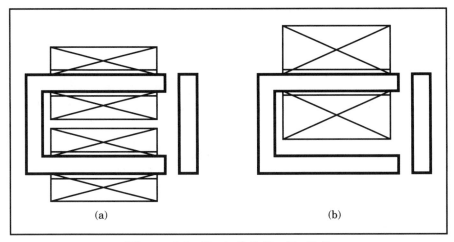

(a)   (b)

Figure 6-9   Single Coil, Double Coil

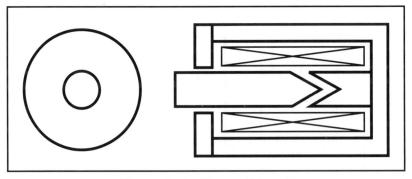

Figure 6-10   Plunger Solenoid

6.8 ELECTRO-MAGNET CONFIGURATIONS

Both DC and AC electro-magnets can be made with linear motion armatures or with pivoted armatures. The pivoted armatures can pivot near one pole, which then does not exert torque but has a low magnetic reluctance to the armature, or can rotate toward both poles together. Linear motion armatures can be either flat-face or plunger type, Fig. 6-10. Different shapes of the plunger ends and matching core projection produce different force-stroke curves.

A rotary configuration is shown in Fig. 6-11. The armature does not approach the pole but varies its overlap over the pole; the effect on magnetic circuit reluctance is identical. If symmetrical, as shown, the radial forces cancel and the rotary forces—the torques—add.

A commercially successful electro-magnet with rotary output is shown in Fig. 6-12. The linear motion of the electro-magnet is converted to rotary motion by the balls rolling along helical grooves.

Flat-face lifting magnets, Fig. 6-7, are DC electro-magnets lowered into contact with their load, then turned on, then lifted. In effect, they are magnetic crane hooks. When the current is turned off, the load is dropped. Large magnets are used to handle scrap metal; small magnets are used to carry control rods in nuclear reactors. Upon either a *scram* signal or a power failure, the control rods are dropped to their bottom position, turning off the chain reaction.

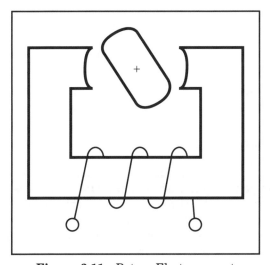

Figure 6-11   Rotary Electromagnet

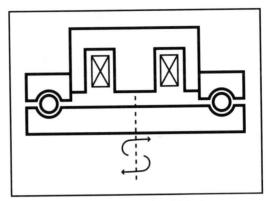

Figure 6-12   Helical Motion

6.9 SOLENOID VALVES

Servo valves using polarized electro-magnets were described above. Two-position valves usually use nonpolarized DC electro-magnets of the plunger configuration of Fig. 6-10. Typically, a fixed sleeve separates the coil and the plunger to prevent fluid leakage, or an insulated coil is in the fluid and is connected electrically through sealed penetrators.

6.10 MOTOR-DRIVEN ACTUATORS

The EM equivalent of a fluid power linear motion cylinder is an electric motor driving either a lead screw, a rack and pinion, or a belt or chain. The motion can be between two end positions equipped with limit switches, or the motion can be subject to transducer feedback and go to any position commanded within the limits of travel. Motor-driven actuators are common industrial products with a wide range of load capacities, travel distances, end fittings, motor types, limit switches, and transducer types.

Rotary actuators are the same except that output motion is rotary instead of linear and a gearbox replaces the lead screw.

6.11 INERTIA WHEELS FOR SATELLITES

Satellites in orbit and deep-space probes must be oriented to point their antennas, solar panels, and instruments. One technique is to put small flywheels on motors on the three axes of the vehicle. When a flywheel— or *inertia wheel*—is accelerated in one direction, the reaction torque accelerates the vehicle in the opposite direction. Thus a change in the inertia wheel's position causes a change in the vehicle's position, and angular momentum of the inertia wheel absorbs angular momentum of the satellite. Steady torque from solar wind or gravity gradient or other cause can accumulate angular momentum in the system. When an inertia wheel approaches its speed limit, a small rocket discharges its angular momentum into space.

6.12 TORQUE COILS FOR SATELLITES

Earth satellites always lie in the earth's magnetic field, but the angle between the earth field vector and the satellite axis may vary from instant to instant, depending on the orientation and shape of the satellite orbit. A set of three energized coils, with or without iron cores, can generate a torque between the satellite and the earth about any axis except the earth's field axis. This effect can be used to supplement orientation rockets in unloading angular momentum from inertia wheels. It uses renewable solar energy instead of expendable rocket fuel.

6.13 STABILIZING GYROSCOPES FOR SHIPS

As an experiment, very large gyroscopes were mounted inside a ship's hull in gimbals pivoted on the pitch axis. When the ship was torqued in roll by the sea, the gyroscopes precessed in pitch and resisted roll. The technique worked but was bulky and expensive. It was replaced by water ailerons extending from the hull and torquing the ship against roll just as ailerons torque airplanes.

6.14 PIEZOELECTRIC ACTUATORS

The piezoelectric effect was described in Section 1.6. Piezoelectric crystals or ceramic bodies are actuators for devices requiring only small displacements at high frequencies. Among these devices are sonar transmitters. In sonars, the actuating piezoelectric can also serve as a sensor. Further discussion of different kinds of sonar is in Chapter 12, "Sound."

Piezoelectric actuators are used to position the *active optics,* the small segments of the large mirror, of the Keck astronomical telescope in Hawaii. The continuous-position modulation is to compensate the varying refraction of the atmosphere.

Some earphones use piezoelectric actuators.

6.15 ELECTRO-STATIC ACTUATION

6.15.1 Dust Precipitators

Electro-static force is used to draw electrically charged dust particles to collecting surfaces in dust removers. The original device, still used, is the Cottrell precipitator. (It was the Cottrell precipitator that cleaned up

Pittsburgh, whose air was once notorious for its burden of dust from the steel industry.) Dusty air (or furnace flue gas) first flows through a high-voltage corona discharge from a set of parallel wires; there the dust particles acquire an electric charge. The dusty air then passes between parallel plates alternately connected to high voltage and ground. The dust particles are driven by electro-static force to impinge on one set of plates, where they remain until removed mechanically.

Large precipitators are mounted on the roofs of power plants to clean their flue gas before it is discharged up the stacks. Small precipitators are used in private houses.

A version of the electro-static precipitator uses friction with the flowing air to generate the high voltage.

6.15.2 Electro-static Loudspeaker

Electro-static force between a metalized flexible sheet and a fixed metal plate generates sound in a loudspeaker. See Chapter 12, "Sound," for details.

6.16 MAGNETIC REPULSION FORMING

When a voltage pulse is applied to a coil, it causes current to rise in the coil. The rising current produces a rising magnetic field and thus a rising flux linkage in the coil. If the magnetic field also links a second coil consisting of a short-circuited turn, it induces a secondary voltage in that turn. The secondary voltage causes a current that opposes the rising magnetic field and thus the rising flux linkage. The secondary current reacts with the rising magnetic field to produce a mechanical force in such a direction as to reduce the opposition to the rising flux linkage. The effect is that the initial voltage pulse drives apart the coil and the short-circuited turn.

If the voltage pulse is the discharge of a large capacitor, the forces can be large enough to squeeze a metal ring into a smaller diameter. If the short-circuited turn starts as a loose-fitting ring around a knob-shaped member, the squeeze can shrink the ring into a tight fit around the knob, similar, in effect, to the mechanical rotary hammering process called swaging. Magnetic repulsion forming is now a commercial manufacturing process.

6.17 MAGNETIC RAIL GUNS

Imagine the short-circuited turn in the preceding section to be a pair of parallel rails, connected at one end, and bridged by a sliding metal dart. The force that prevents opposition to rising flux linkage urges the dart along the rails, thus increasing the area enclosed by the short-circuited turn and through which the magnetic field must pass. Tom Swift's science fiction electric cannon has come to pass. Actual numbers achieved in experimental apparatus were velocities higher than can be achieved with gunpowder. (The objective was a "Star Wars" weapon.)

6.18 MAGNETOSTRICTION ACTUATORS

Magnetostriction is a dual of piezoelectricity: A magnetic field causes magnetostrictive material such as nickel to constrict.

There was described, earlier in this book, an actuator used to precisely adjust large, centerless grinders. A long nickel bar, 2 inches in diameter, was surrounded by a coil. A bar clamp capable of being switched on and off was placed at each end of the coil. The coil and clamps were energized cyclically so the bar moved forward or backward, depending on the sequence, in the manner of an inchworm; that resemblance provided its trade name [A34].

6.19 MICROMINIATURE ACTUATORS

Actuators are being made of etched silicon using similar techniques to those used to make integrated circuits and silicon transducers. These consist of electro-static motors and a variety of devices based on the bending of tiny beams [A53].

CHAPTER 7

Transducers

7.1 GENERAL PRINCIPLES

This chapter starts with some general principles of sensing instruments, then lists phenomena that can be sensed by instruments, and then describes specific devices to illustrate the principles. It does not catalog the very large number of transducers that exist, except to identify some references that do.

Transducer is a pedantic word for measuring instrument. "Trans" was derived from the Latin for across, and "duce" was derived from the Latin for lead; a transducer leads a physical quantity across from one form, such as hydraulic pressure, to another form, such as electric voltage.

Usage limits the word to automatic instruments, that is, those not requiring a human operator. However, it may generate a deeper understanding to first consider some simple instruments not usually classified as "transducers." For example:

1. A 12-inch desk ruler *transduces* a physical length to a displayed number on a scale, for example, 5¾.

2. A mercury thermometer transduces a temperature to a displayed number on a scale.

3. A voltmeter transduces an electric voltage to a displayed number on a scale. (This example contradicts the usual assumption that a transducer generates an electrical output from a mechanical input.)

4. A current transformer transduces a large electric current flowing through a conductor of large cross section in one or a few turns to a small, proportional, electric current flowing through a conductor of small cross section and many turns.

5. A pH electrode transduces the acidity of a solution to an electric voltage.

Other transducers, not electro-mechanical, are:

1. A thermocouple transduces temperature to voltage.

2. A photoelectric cell transduces light to electricity. (More exactly, a photovoltaic cell generates electric voltage and a photoresistive cell varies its electrical resistance as a function of its illumination.)

3. Many transducers convert physical quantities into pneumatic pressure. This pressure is transmitted through tubing, signals pressure balancing *pneu-matic relays,* and controls pneu-maticmechanical devices, particularly valve actuators.

The reverse of a *transducer* is an *actuator* that converts power in one form, such as voltage, to action in another form, such as motion. An electro-magnet is such an actuator. Chapter 6 dealt with actuators.

Electro-mechanical transducers transmit *information* about the mechanical world into the electrical world just as electro-mechanical generators transmit *power* from the mechanical world into the electrical world.

The difference between information and power is function and magnitude; most information transmission sends a small amount of power, and there certainly is information in whether or not a generator is running.

Sonar actuators that transmit acoustic power into water are called transducers rather than actuators or transmitters. Similarly, medical and other ultrasonic actuators are also called transducers. Many ultrasonic transducers both transmit and receive.

Radio transmitters are also actuators converting circuit power into radiation power, but they are not called actuators except in philosophical essays like this one. Usually, an actuator is thought to have a limited range, like a solenoid, while a transmitter or power converter operates without time or motion limits, like a motor.)

A transducer's information may have the range of only one binary digit, as in a limit switch, or it may have the highest accuracy and range man can create, as in an inertial guidance accelerometer.

Not only do EM transducers measure and *telemeter* mechanical parameters via electricity, but some mechanical measurements are most easily and accurately made by EM instruments. Usually the mechanical parameter imparts a small amount of energy to the transducer, the transducer translates the magnitude of that energy into an electrical form, and that magnitude is either displayed, or used in a control circuit, or transmitted.

Some transducers are direct. Examples:

1. A tank-level transducer has a float that moves a potentiometer wiper. The voltage between the potentiometer end and the wiper is the output.

2. A wind-speed transducer has a propeller-driven generator whose voltage is a measure of wind speed. Similar generators are used for machine tachometers. Their outputs are either a voltage or a stream of pulses whose rate is proportional to speed.

Some transducers are non-linear, or are ambient-temperature sensitive, or may have very low voltage or power outputs. Therefore, they may have signal conditioning circuitry included in their packages. Such circuitry may include amplification, filtering, non-linearity correction, and analog-to-digital conversion.

Some transducers are null instruments; that is, either the original input or the sensing element output is balanced by feedback [2.13]. The final transducer output is the output of the feedback circuit. Examples:

1. Most of the inertial guidance transducers described below (gyroscopes and accelerometers) are null instruments.

2. Very accurate measurements of linear and angular dimensions are made by mounting the workpiece on the equivalent of a large machine tool called a measuring machine. The measuring machine moves the workpiece until a sensitive EM position sensor touches a surface and reads null. The dimensions are read from the calibrated drives of the measuring machine or from position transducers on the outputs of those drives. The null sensor may be used in a feedback loop to drive the measuring machine to the null position.

7.2 INTERMEDIATE PARAMETERS

Many EM transducers generate an intermediate parameter, such as mechanical displacement. (The mechanical displacement may be material strain [Section 2-3].) The intermediate parameter is then sensed by another transducer, such as a strain gauge, that generates the output electrical signal. Examples:

1. Fluid pressure or vacuum stresses a diaphragm, and the diaphragm's strain is sensed by a strain gauge.
2. A force stresses a part whose strain or bending displacement is sensed by a strain gauge.

3. Float motion displaces a potentiometer, as above.

4. Bimetal deformation or fluid expansion from temperature change generates a displacement.

Some transducers inject power into the system being measured and then sense certain effects of that power. Examples:

1. Radar and sonar radiate radio power and sound power, respectively, toward a target and measure distance by measuring the time for an echo to return.

2. Electrical parameters (*R, L, C,* and so forth) are measured by their effect on electricity injected into circuits containing the parameters, for example, a Wheatstone bridge.

3. Light is injected into systems measured by their effect on that light, such as pulsing it on and off.

Other forms of power used in a transducer system are heat, pressure, and force.

7.3 PARAMETERS SENSED

The parameters with which electro-mechanical transducers are used include the following:

1. Force (from dynes to thousands of tons)

2. Pressure (from high vacuum to tens of thousand of pounds per square inch)

3. Temperature (from absolute zero to the temperature of the sun)

4. Linear displacement (from fractions of a light wave to miles)

5. *Proximity,* the presence of an object close to the transducer

6. Angular displacement (from microradians to unlimited revolutions)

7. Frequency (from zero to gigahertz)

8. Time (as the count of oscillation cycles)

9. Viscosity

10. Vibration, including sound (frequency, amplitude, harmonics, direction)

11. Motion in space (Acceleration, linear and angular. Velocity, as integrated acceleration. Position, as integrated velocity. These are the parameters of inertial guidance.)

12. Flow (gases and liquids)

13. Chemical composition

7.4 ELECTRICAL SIGNAL GENERATION

The purpose of this book is understanding, and the purpose of this chapter is understanding transducers. Therefore, the following examples are classified by transducer principle, that is, by the means used by the transducer to generate electrical signals rather than by the parameters measured. A systems engineer doing a design would start with the parameter measured and then find an appropriate transducer.

First is an overview of the phenomena used to generate electrical signals, then, in the same order, are examples of transducers using these phenomena.

1. Resistance change
2. Self-inductance change
3. Mutual-inductance change
4. Capacitance change
5. Frequency
6. Tuning sharpness
7. Acoustic effects
8. Generated voltage
9. Serial code (pulses)
10. Parallel code
11. Voltage ratios (synchros)
12. Inertial effects (accelerometers and gyros)
13. Electro-magnetic radiation (EMR) effects

7.4.1 Resistance Change

Closing or opening a contact switches resistance between zero and infinity. The information is only one binary digit, but it may be sufficient, and the technique is simple and inexpensive. The contact may be in a mechanically operated switch such as a *limit switch,* a magnetically operated switch such as a *reed switch* operated by a permanent magnet, or a liquid-level touch electrode. Certain limit switches used as touch detectors in mechanical gauging systems have a sensitivity of 1 micron.

Limit switches are binary transducers to indicate when a movable member is at one side or another of a particular position on its path. Typical movable members are doors and machine slides.

Limit switches for machine tool and other rough service are made with great abuse resistance. The movable member may be a rotating lever or a sliding plunger. It may be returned by an internal spring or by reverse motion of the machine. Mechanical contact with the sensed machine is usually via a roller on the tip of the lever or plunger to reduce friction. Other movable members are flexible *whiskers,* double levers, and so forth.

There is a class of small snap-acting switches first made by the Micro-Switch Division of Honeywell, Inc. and commonly called *microswitches,* although "Micro-Switch" is a trade name owned by Honeywell, Inc. The snap action prevents arcing at the contacts when the switch is operated slowly. Typically, the displacement of the operating plunger between the open and closed positions is only a few thousandths of an inch. Their manufacturers offer a variety of lever and roller attachments to adapt the switches to different operating geometries. Some snap-acting switches are hermetically sealed and are operated through a flexible diaphragm in the enclosure; others are usually in some form of enclosure to protect them from their environments.

Small mercury switches are used as limit switches, actuated by tipping the switch through a substantial angle. Mercury switches are insensitive to the environment, have very long lives, and are highly reliable.

Reed switches [Chapter 9] are used as limit switches. They are operated by a permanent magnet moved near or far, so closing and opening geometry is not accurately repeatable. They are small and insensitive to their environment but have very low current ratings. Many are used in burglar alarms.

Limit switches are sometimes replaced by proximity transducers, described below. They use a variety of effects but end up with a binary signal that a target either is or is not within their sensitive range.

A class of limit switch is the revolution counter. The controlled machine rotates a shaft through a speed reducer (gears or chain and sprocket.) One or more adjustable cams on the shaft operate and release snap-acting switches, commanding a program.

Devices that vary the magnitude of resistance are:

1. Sliding the wiper of a rotary or linear potentiometer changes the resistances between the wiper and the ends. This technique requires that input displacement be as large as the resistor element and that motion frequency be not so great as to wear out the resistor too soon. A potentiometer can handle relatively large voltages and currents so no sensitive amplifier is needed.

There can be mechanical scaling and conversion before the motion reaches the potentiometer. For example, a *lanyard transducer* has a long cord wound on a capstan geared to a potentiometer. A spring maintains cord tension.

2. A non-linear resistor element produces a non-linear output, which may be useful. A special case uses a square resistor winding and two pairs of brushes. One pair produces a sine function of angle while the other pair produces a cosine function.

3. Stretching or compressing an electrical resistor changes its resistance. Such *piezo-resistive* resistors are used as *strain gauges*. The resistors are often connected in a bridge circuit. Adjacent calibrating resistors may be made as thick film deposits that are then laser trimmed. Sometimes strain gauge resistors are arrayed to sense strain on only one axis, and sometimes a set of strain gauge resistors are arrayed in different directions so the direction of the strain can be measured.

Bonded strain gauges are made of resistors adhesively attached to an insulating substrate that may, in turn, be adhesively attached to the member whose strain is to be measured.

Some strain gauges are made by semiconductor techniques on the surface of a silicon chip. The chip also contains conductors that connect the resistors to circuitry on the same chip. On silicon substrates the strain gauge resistors are made by ion implantation.

The chip is sometimes micromachined to be an elastic diaphragm to be deformed by either fluid pressure or acceleration.

These devices are both miniature and inexpensive [A53].

Unbonded strain gauges are wound from wire onto insulating supports that in turn are supported by the member whose strain is to be measured. This member may be part of a machine or structure, or it may be a block of metal so sized that it reaches its greatest allowable strain at the greatest load to be measured (a *load cell*), or it may be a diaphragm deformed by fluid pressure. Different load cells are made to transmit load in tension, or compression, or torque.

In a carbon granule microphone, sound pressure on a diaphragm varies resistance by varying the force between particles of carbon.

7.4.2 Self-Inductance Change

Varying the air gap in the magnetic circuit of an iron-core inductance changes the self-inductance. If the air gaps of two inductances are changed by the same motion, one increasing when the other is decreasing, the device is a displacement null detector. (Null is when the inductances are equal as detected by a bridge circuit.)

Varying the space between an iron part and a sensing coil is the basis of one kind of proximity sensor.

Varying the space between a conducting surface and a sensing coil changes the impedance of the coil. The conductor acts as a shield to the AC field of the coil and affects the net permeance of the magnetic circuit. This is the basis of a proximity sensor for non-magnetic parts.

7.4.3 Mutual-Inductance Change

Change in the mutual inductances of two or more coils can produce a signal voltage. For example, Fig. 7-1 shows a differential transformer having primary coil P inductively coupled to two secondary coils S1,S2

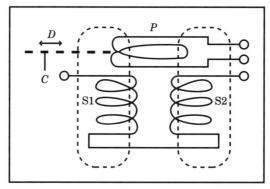

Figure 7-1   Differential Transformer

connected in series. When there is symmetry, the two secondary voltages are equal and opposite and the output voltage is zero. If dissymmetry D is introduced by motion of mechanical coupling C, one secondary voltage increases, the other decreases, and an output voltage appears. Its magnitude varies directly with the magnitude of D, and its phase corresponds to the direction of D.

Differential transformers may be designed for linear or rotary displacement. They can be made with microinch sensitivity for extremely precise mechanical measurement, with microradian sensitivity to measure gyroscope displacement, or with working ranges of many inches or degrees.

An unusual application of this principle was a linear differential transformer to position a crawling machine inside a steel pipeline. The primary was an iron-core winding manually positioned near a weld on the outside of the pipe. Enough flux penetrated the steel to be sensed by the secondaries on the machine. When their voltages balanced, the machine automatically stopped and uncovered a radioactive source to X-ray the weld.

Dissymmetry can be generated by relative coil motion, as in Fig. 7-1 or with fixed coils and a moving ferromagnetic core, as in Fig. 7-2.

The coils may be arrayed for linear displacement, as suggested by Fig. 7-1, or divided into two or more sets arrayed around a circle for angular displacement.

The transformer may be entirely air core, Fig. 7-1, or it may have a single iron structure carrying the magnetic flux of the primary coil, Fig. 7-3, or it may have separate iron cores with the flux passing from iron to iron across small gaps, Fig. 7-4.

The two-core design is the most sensitive because the total magnetic reluctance is least and a small displacement makes the greatest change in flux division. However, there are magnetic forces between the two

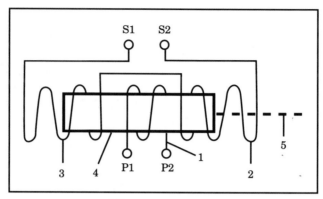

Figure 7-2   LVDT

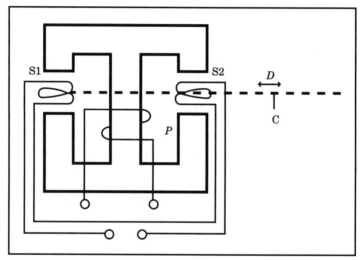

Figure 7-3   Differential Transformer, Single-Iron Core

pieces. When measuring the displacement of a sensitive device such as a gyroscope, these forces can be balanced only to the degree that manufacturing tolerances approach zero; symmetry error introduces torque error.

Figure 7-2 shows a *linear voltage differential transformer* (LVDT). It consists of a primary coil 1 connected to input terminals P1,P2 and two secondary coils 2,3 connected in series, so that their induced voltages oppose each other, and connected to output terminals S1,S2. A movable magnetic core 4 is positioned in the center but is movable axially via member 5 coupled to the element whose position is to be sensed. When an AC voltage is applied to P1,P2 voltages are induced in secondary coils 2,3. If core 4 is in the center, the two secondary voltages are equal and opposite,

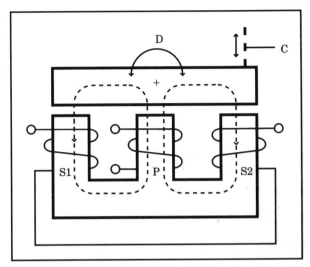

Figure 7-4   Differential Transformer, Two-Iron Cores

so no voltage appears on output terminals S1,S2. If the core is moved off center, the magnetic field from primary coil 1 shifts with it, and one secondary voltage increases and the other decreases. A net voltage appears on terminals S1,S2. Its magnitude varies with the amount of core displacement, and its phase reverses as the direction of core displacement reverses.

LVDTs are made with resolutions less than 0.0001 inch and with full-scale displacements up to many inches.

7.4.4 Capacitance Change

The capacitance between a fixed plate and a moving or vibrating diaphragm comprise a microphone. The capacitance varies with the spacing and is measured with a high-frequency circuit. A pair of fixed plates provides a differential capacitor, Fig. 7-5.

The capacitance between two fixed plates with a dielectric liquid between them varies with the level of the liquid between the plates; the capacitance is a measure of liquid level. Similarly, the introduction of a solid dielectric in the electric field between two plates changes their capacitance; this is the basis of another form of proximity sensor.

7.4.5 Frequency

The natural frequency of oscillation of a string depends on the tension in the string. In a kind of load cell, the string is a wire between magnetic poles and is part of an oscillator whose frequency is the string's natural frequency. The tension may come from a pressure-sensing diaphragm or

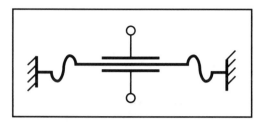

Figure 7-5   Differential Capacitor

it may be the acceleration force on a proof mass in an accelerometer. The information output is the oscillator frequency. Frequency as a quantity representative is immune to noise.

7.4.6 Tuning Sharpness (Q)

Q is the symbol for sharpness of tuning in a resonant system; it represents the smallness of damping. An example of its use is in measuring viscosity by inserting a mechanically oscillating member into a viscous fluid and measuring the Q of the oscillator.

Yet another proximity sensor comprises a coil in an oscillator circuit. When the coil approaches either magnetic or non-magnetic metal that dissipates power from the coil's magnetic field, the circuit's Q drops.

7.4.7 Acoustic Effects

Since sound is mechanical vibration, devices combining sound and electricity are electro-mechanical.

See [7.4.1] above for the carbon granule microphone.

A piezoelectric crystal coupled to either a microphone diaphragm or a phonograph needle generates an electrical voltage corresponding to the acoustic input to the diaphragm or needle.

A piezoelectric crystal on which a small mass is mounted generates a voltage proportional to the acceleration force on that mass. This effect is used in an accelerometer and vibration sensor.

Still another proximity sensor is a sonar-in-air device that emits an ultrasonic beam and signals when it receives a reflection.

A distance-measuring transducer transmits an ultrasonic signal along a magnetic steel rod. A moving coil surrounding the rod receives the signal and returns it. (The coil and its circuitry are a *transponder*.) The round-trip time is measured as a measure of distance from the transmitter/receiver to the transponder.

Underwater sonar is used for underwater detection, ranging, and direction finding. The targets are submarines, fish, and the bottom.

Diagnostic sonar is used to observe structures within the human body.

Doppler sonar is used to measure fluid flow, including non-invasive blood flow measurement in the human body.

Sonar is used in geophysical exploration to sense and measure the positions and shapes of layers of different materials in the earth, in materials testing to detect flaws in the material, and in laminate testing to detect unbonded areas. (Here the word *sonar* means all forms of sonic and ultrasonic sensing of reflected sound.)

Medical and engineering stethoscopes may transmit sound generated within a person or a machine directly to the ear or may amplify it with an EM microphone and electronic amplifier.

Microphones in sound systems use electro-dynamic, piezoelectric, or capacitive effects. They sense sound in the air, in water, and in solids.

See [Section 2.1 and Chapter 12, "Sound"].

7.4.8 Generated Voltage

Figure 7-6 shows an electro-dynamic microphone. Permanent magnet 1 establishes a radial magnetic field to soft iron yoke 2. Coil 3 on spool 4 lies in the field, and a voltage is induced in coil 3 proportional to the speed of its motion; the motion is transmitted by coupling 5 from the element whose motion is being sensed.

This microphone is the same as a loudspeaker (or earphone) except that its motion is used to generate a voltage whereas in a loudspeaker, the coil current is used to generate motion.

A stereo phonograph cartridge for grooved records has two microphones, each driven by one component of motion of a single needle. The microphones may be either piezoelectric or electro-dynamic.

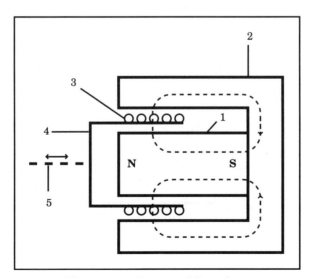

Figure 7-6 Dynamic Microphone

Hall effect, See Fig. 7-7. The effect resembles, in a solid, the bending of an electron beam by a magnetic field in a cathode-ray tube. A steady current flows through resistive member 1 between terminals 2,3. Normally no differential voltage appears between contacts 5,6 but if a magnetic field 4 passes through member 1, the current path is deformed and voltage appears between terminals 5,6.

A piezoelectric crystal generates a voltage when stressed by a mechanical force.

A coil in a pulsating magnetic field produces an electrical pulse for each magnetic field pulse. Figure 7-8 shows a fixed coil 1 and a fixed permanent magnet 2. Iron gear teeth 3 move past the magnet and cause its field strength to pulsate. The electric pulse rate is a measure of the gear's speed.

A rotating generator produces a voltage proportional to its rotation speed; this is the most common kind of tachometer. If rotation is produced by a propeller in a flowing fluid, the transducer is a flowmeter.

A flowmeter for conducting liquids that introduces no moving parts into the liquid is made by using a non-conducting pipe, a pair of electrodes at the ends of one diameter, and a magnetic field along a perpendicular diameter. A voltage is generated by the conducting liquid flowing across the magnetic field and is sensed by the electrodes. The voltage is proportional to the flow speed.

7.4.9 Serial Code (Pulses)

A simple example of a pulse-counting transducer is the photoelectric drop counter in a medical intravenous (IV) supply, Fig. 7-9. Fluid from container 1 is *metered* by valve 2, passes as a series of drops 3 through

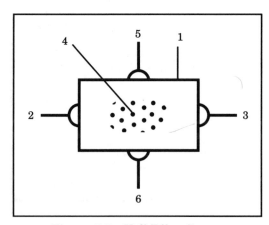

Figure 7-7   Hall Effect Sensor

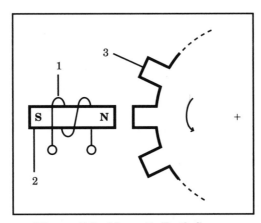

Figure 7-8   Magnetic Tooth Sensor

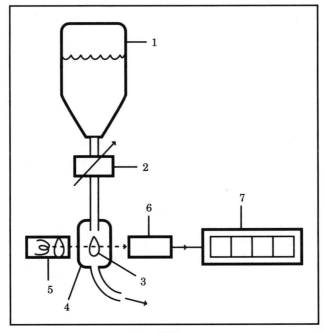

Figure 7-9   IV Drop Counter

transparent drop counter 4, and thence flows to the patient. Each drop momentarily interrupts light from lamp 5 to photocell 6, and the interruptions are counted and displayed by electronic assembly 7. The display shows the total number of drops and therefore the total quantity of liquid that has passed. If a timer is included, the display can show flow rate.

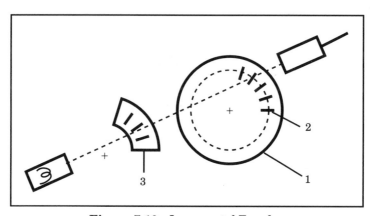

Figure 7-10   Incremental Encoder

Flow is always one way so there is no problem of sign. The accuracy depends on the uniformity of drop volume.

An *incremental encoder,* Fig. 7-10, senses shaft rotation by providing, on the shaft, a transparent disc 1 having a ring of opaque radial lines 2. A lamp and photocell count the lines as the disc rotates and the lines pass between the lamp and the photocell. The signal is stronger if the lines and spaces are of equal width and if a stationary mask 3, having a matching set of lines and spaces, is placed in the light path. Then the light from many spaces is sensed and there is no need to focus the light onto a single line width. The pulse frequency is a measure of shaft speed, so the same transducer, with different circuitry, is also a *tachometer.* Sometimes, an additional opaque concentric ring is provided with a single transparent line. With a separate lamp and photocell, it serves as a revolution counter that can verify the accuracy of the main count.

If the shaft sometimes reverses, it is necessary to sense the sign of rotation to determine the position of the shaft. In Fig. 7-11, disc 1 is provided with three sets of lamps, masks, and photocells, A,B,C. The masks 2 are positioned so that their photocells 3 are illuminated in sequence as the shaft rotates. If the sequence is ABCABC, rotation is in one direction; if the sequence is ACBACB, rotation is in the other direction.

With linear motion, there is usually a need to sense direction since most linear motion devices must reverse. Furthermore, the principal application is the measurement of machine tool slide motion with a resolution of 0.0001 inch, so mask positioning is difficult. The usual solution uses a skewed mask as in Fig. 7-12. Moving scale 1 is marked with lines spaced 0.0001 inch apart. Fixed mask 2 has a similar line pattern, but its pattern is at a small angle to the moving scale pattern. One side of the scales is illuminated, and there are three photocells A,B,C on the other side. The transmitted light forms wide bands of light and dark, *moiré*

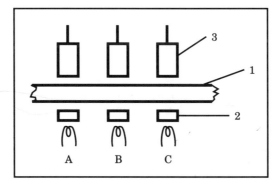

Figure 7-11   Rotation Sign (CW or CCW)

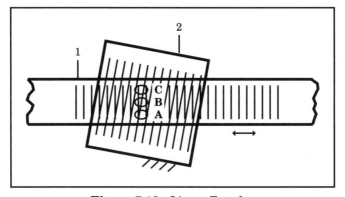

Figure 7-12   Linear Encoder

fringes, parallel to the moving scale. These bands move perpendicular to the moving scale, one cycle per scale line of motion. The result is alternating illumination of each photocell in sequence ABCABC or ACBACB, depending on the direction of motion.

The same moiré fringe counting system is used in interferometer measurement of linear motion except that, instead of counting scale graduations, the photocells count light waves. Laser light is divided between a fixed path and the linear motion path and is reflected by mirrors at the ends of each. The reflected light waves interfere with each other to form light and dark bands that move across the photocells.

Bar code identification and bar code readers are another case of serial code. We are all familiar with food market identification bar codes and their laser readers at checkout counters. Bar codes differ from simple pulse codes in that bar code pulses are of different duration, rather like Morse code. The laser spot scans via motor-driven rotating mirrors. Bar codes are also used for railroad car identification, in some postal systems, and in industrial warehousing automatic storage and retrieval systems.

Magnetic stripes and stripe readers are used to identify credit cards, identification cards, and certain train fare and telephone charge tickets. A serial code is written on the magnetic stripe in the form of magnetized cross stripes. The stripe is passed across a magnetic sensing head, inducing a pulse train corresponding to the magnetic code. In train fare and pay phone systems, the user pays for a new card when the code reads full. At each pass through a reader, a computer subtracts some money and rerecords the balance.

7.4.10 Parallel Code

All the position-sensing systems just mentioned are *incremental;* they count increments of position rather than sense *absolute* position. Absolute systems generate a simultaneous multibit code that defines the actual position of a member even when that member is not moving.

Instead of a single row of identical marks, an *encoder* has a separate concentric ring, or parallel stripe, for each of the binary digits required to define a position. Each ring or stripe has marks whose length corresponds to the length of travel corresponding to its bit magnitude.

The marks may be opaque or transparent regions with photoelectric sensors, or conducting or non-conducting regions with conducting brushes as sensors, or AC windings in slots between magnetic teeth with similar windings, slots, and teeth as sensors.

Optical character recognition (OCR) systems use either moving mirrors or video cameras to present printed characters, one at a time, to a rectangular array of photodetectors. The pattern from illuminated or dark detectors is compared with the memorized patterns of each character in a font. The output of each photodetector (or video bit) corresponds to gray rather than to black or white because of imperfect printing and dirt. The circuit measures the grayness at each detector and chooses the memory character with the best match. The reading of each character is by electronics, but the sequencing of characters, or of character lines, is by EM motion of the paper or mirrors.

Bank check code uses a special font printed with magnetic ink. The line of characters is scanned serially with a cross row of magnetic heads reading simultaneously.

7.4.11 Voltage Ratios (Synchros)

A class of motor-like devices is used to transmit angular position over electric wires. The translation of angular position to electrical signals is done by changes in the mutual inductance between the rotor and stator coils of the device.

Figure 7-13 shows two identical electrical machines A, B. Each has a rotor winding 1,1' energized from the same single-phase source. Each has three stator windings 2,3,4 and 2',3',4' spaced 120° apart as in three-phase

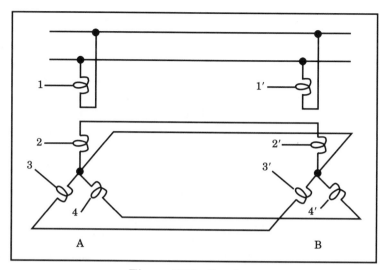

Figure 7-13   Synchros

power generators or motors. *Single-phase* (that is, the same single *time* phase) voltages are induced in the three stator windings despite their three separated *space* phases. The *magnitude* of each stator voltage depends on the mechanical angle between that stator winding and the rotor winding. If the rotor positions are the same in each machine, the stator voltages match, and there is no current flow in the stators and no torque. If the rotor angles are different, the voltages do not match, current flows, and torques appear in both machines in the direction of restoring matching positions. The coupling is elastic rather than rigid because the torquing current is proportional to the position error.

In a variation of this scheme, the synchro chosen as the receiver has its rotor winding turned 90° and connected to an amplifier input instead of to the AC supply. Stator currents flow, and a single-phase voltage is induced in the receiver rotor corresponding to the angular difference between machines beyond the original 90°. Its phase reverses with the sign of that difference. The rotor voltage is amplified and drives a servomotor that turns the receiver synchro to null, and with it, a load that may be as large as desired. The synchros can be quite small yet still control the position of heavy loads.

For extreme accuracy, two sets of synchros are geared together as a coarse and fine system, typically at a ratio of 36:1.

7.4.12 Inertial Effects (Governors, Gyros, and Accelerometers)

The world's first transducer used in a feedback control system was the centrifugal speed governor; and it is still in use today. It was invented by James Watt to maintain the speed of his steam engines. Figure 7-14 shows a vertical shaft 1 rotated by the engine output. Arms 2 are pivoted to shaft 1 and carry flyweights 3. At any speed, there is an equilibrium position of the flyweights in which gravity-restoring torque balances centrifugal force. Links 4 transmit flyweight position and lift rotating *shifter yoke* 5 to a corresponding position. Rotating shifter yoke 5 moves nonrotating throttle lever 6. This motion feeds back the mechanical transducer signal to the steam throttle valve that is the amplifier in the feedback control loop.

We now enter the world of electro-mechanical transducers of the highest possible precision: the world of inertial guidance. Airplanes, ships, submarines, torpedoes, and short- and long-range missiles use inertial guidance for automatic control of their paths.

Gyroscope precession is explained in Section 2.5.2. The *single-axis gyro* in Fig. 2-3 ignores rotation of gimbal 6 around the Z-axis (its spin axis), it precesses around the Y-axis at a rate proportional to the torque applied to it around the X-axis via gimbals 6 and 3. Rotation of gimbal 6 around the Y-axis leaves the wheel orientation unchanged.

An accelerometer has a *proof mass,* or *seismic mass,* that is subject to a force if the accelerometer is accelerated along its sensitive axis. Most are null instruments in feedback control loops that prevent more than very small actual deflections. Since most of their control systems use digital computers, the feedback usually starts as a train of pulses and integration is done by counting pulses.

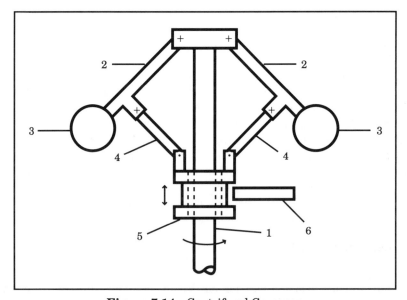

Figure 7-14   Centrifugal Governor

In practice, each device is made sensitive to motion on only one axis because it can be made more accurate that way; therefore, both gyros and accelerometers are used in sets of three.

There are many variations of both devices and of systems using both devices [A18, A23, A26, A27, A28, A30]. One of particular interest in the study of EMDs is the use of a gyroscope as an integrator of an accelerometer output. The accelerometer output is a torque on the gyro and the gyro's accumulated precession angle is proportional to the integral of the accelerometer output.

7.5 TRANSDUCER ACCURACY AND ERROR ANALYSIS

Error analysis is fundamental to the design of any instrument; an understanding of the sources of error in electro-mechanical instruments gives understanding of the nature of electro-mechanical engineering. The extreme accuracy required of inertial guidance instruments drives the study of error sources to its limit. Among the sources of error considered are the following:

1. Anisoelasticity of materials
2. Anorthoganality of manufacture
3. Eccentricity of center of mass to center of buoyancy
4. Convection currents in buoyancy liquids
5. Temperature gradients and excessive temperature
6. Dimensional stability of materials
7. Uniformity and constancy of electrical pulses
8. Magnetic circuit symmetry
9. Static and dynamic unbalance of the rotor
10. Cross coupling of forces and deflections on orthogonal axes
11. Elastic deformation under stress
12. Forces from stray magnetic fields
13. Manufacturing tolerances in parts and in assemblies
14. Microdamage during assembly
15. Slow chemical reactions (for example, corrosion)
16. Wear, particularly in bearings
17. Looseness: threads, pivots, and so forth
18. Entrance of contaminants during manufacture or during service
19. Inadequate cleaning during manufacture (for example, solder and adhesive joints)
20. Abuse during transport, storage, or use (mechanical, electrical, magnetic)

21. Overload, electrical or mechanical

22. Decay of permanent magnets

All electronic components and interconnections are electro-mechanical structures designed and used for their electrical properties. Usually, there are no moving parts if they are considered to be "electronic." Failures in electronic circuits, whether infant mortality or long term, come from the same list of sources as the failures in electro-mechanical and all-mechanical devices.

Because of the delayed appearance of latent defects, assuring service within specifications can require far more than inspection of the product. It is necessary to assure proper design, proper materials, and proper manufacturing processes, including storage, handling, and shipment. Furthermore, some testing requires destruction or aging, so testing samples to destruction, with statistical quality control, may be necessary. (Consider the melting current of a fuse.)

7.6 ELECTRO-MAGNETIC RADIATION (EMR)

Electricity generates EMR by heating filaments and arcs, by bombarding phosphors and other targets with electrons, by solid-state effects in light-emitting diodes, by energizing lasers, and by electronic oscillators.

Conversely, infrared and shorter wavelengths generate electricity by ionizing atoms in both photoconductive and photovoltaic cells; longer waves generate electric signals in a variety of other detectors.

EMR devices measure the wavelength of visible light by varying the path geometry of light passing through refraction prisms and gratings. On a very large scale, synchronized radio telescopes are interferometers for microwave radiation from stars.

Examples of the transducers that use intensity, phase angle, and wavelength of radiation are the following:

1. Optical pyrometers measure temperature.

2. Telescopes' spectrometers measure the speed of stars by measuring the red shift of their spectra.

3. Spectrometers sense the presence of elements in instrument arcs and in stars by measuring their emission spectra.

4. Colorimeters analyze the color of cloth, paint, and the like.

5. Medical X-ray film and fluorescent screens shadow bones and tissues.

6. Chemical analyzers absorb or transmit different wavelengths.

7. X-ray thickness gauges measure the thickness of metal moving through a rolling mill.

8. X-ray diffraction apparatus analyzes crystal structure.

9. Imaging systems such as Magnetic Resonance Imaging (MRI) image soft tissues.

10. Proximity sensors (photoelectric sensors) use transmission, blocking, and reflection of light.

11. The change in transmission by optical fibers under stress is a measure of that stress.

12. Fiber optic and ring-laser gyros are replacing electro-mechanical gyros in guidance systems. A beam of laser light is divided between clockwise and counterclockwise paths, and the two beams form interference patterns when they recombine. Angular velocity of the two paths results in changes in interference patterns.

13. Target trackers, such as radar, use microwave frequencies. Star trackers use visible light. Electro-mechanical antennas and telescopes follow the electronic detector signals to keep the antenna or telescope pointed at the target.

One class of EMR-based transducer is distance-measuring devices. Examples:

1. Triangulation devices such as binocular range finders

2. Laser sources combined with focused photocells sense the presence of an object.

3. Video-imaging gauges sense small variations in distance.

4. Distance measuring by measuring visible-light echo time from a reflector (as in radar)

5. Measuring phase angles of different RF frequencies reflected from a distant object to measure its position and speed.

6. Extremely accurate interferometer measurement of distance by counting wavelengths of light as the distance changes.

7. The Global Positioning System (GPS) uses radio signals from satellites to determine position on Earth.

7.7 NON-ELECTRO-MECHANICAL TRANSDUCERS

7.7.1 General

The simplest transducer is the whistle on a teakettle that transduces steam pressure into acoustic radiation to announce boiling.

There are purely mechanical systems, such as the mechanical linkages in railroad yards, that transmit signal and switch positions to the control tower. The grandfather of all transducers for feedback control systems is the centrifugal governor on a steam engine that directly adjusts the throttle of the engine, as described above. [Section 7.4.12]

7.7.2 Clocks and Timers

Clocks transduce time, mostly into visual displays. *Timers* are clocks that issue commands at programmed instants of time.

Clocks range from ancient water clocks, through calibrated candles, sundials, sandglasses, medieval jackwork displays, pendulum clocks, tower and grandfather clocks, mechanical alarm clocks, wristwatches, electric motor-driven clocks, and quartz and maser digital clocks.

Pendulum, balance wheel, maser, and quartz clocks count the cycles of mechanical oscillators and provide the power losses of those oscillators. (Quartz crystals are mechanical oscillators, piezoelectrically sensed and driven. *Masers* use the mechanical oscillation of atoms in molecules.) Electric circuit oscillators such as inductance-capacitance resonant circuits or resistance-capacitance relaxation oscillators are much less stable and accurate than these mechanical and EM oscillators.

Cycle counting is done by electronic counters and by mechanical and EM counters such as escapements and ratchets with gear trains.

The earth's rotation and its orbit around the sun, observed by astronomical telescopes, has been the ultimate definition and measure of time, but small variations are observed even in them. Certain atomic oscillations are now the standard.

DC phenomena such as water clocks and sandglasses are much less accurate than the oscillators described above. Nevertheless, for much short-period timing, quite useful timers use resistance-capacitance circuits, air flow through orifices, rotating propellers on bomb noses, burning fuses, or heat flow.

Synchronous motor clocks are merely repeaters for the clocks that establish utility system frequencies.

7.7.3 Pneumatic Transducers

Figure 7-15 (a),(b) shows air nozzles used as proximity transducers. In (a) compressed air flows through nozzle 1 and escapes to atmosphere through the clearance between nozzle 1 and vane 2. The pressure drop across the clearance is a measure of the spacing between the nozzle and the vane. Feedback control pneumatic circuits and amplifiers, analogous to electric circuits and amplifiers, sense this pressure drop and respond to it in ways to restore the spacing to its original value or merely to indicate proximity.

Figure 7-15(b) shows a somewhat more sensitive dual nozzle in which air flows only through outer annular nozzle 3 and feedback pressure is sensed by inner nozzle 4 without steady-state flow through it.

Pneumatic control systems are still widely used, particularly in chemical industry, because pneumatic valve actuators are: (1) versatile, relia-

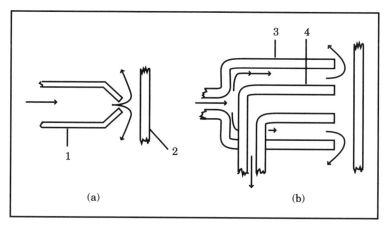

Figure 7-15 Pneumatic Transducers

ble, and inexpensive; (2) inherently cannot start fires and explosions; and (3) are resistant to corrosion and heat.

7.7.4 Thermal Transducers

Heat expands almost all gases, liquids, and solids. That expansion can operate EM displacement or pressure transducers and can directly operate valves.

Temperature-sensing gases and liquids are usually enclosed in expandable bellows.

Solids are usually laminated into *thermostatic bimetal,* two layers of different metals with different coefficients of expansion, diffusion-bonded together. When temperature changes, shear stress at the bond causes the laminate to bend.

Any two wires of different metals connected in a loop at two different junctions generate a net loop voltage if the temperatures are different at the two junctions; the junctions are called *thermocouples.* This voltage is sufficient to operate a galvanometer or to be amplified to control the heat source.

The electrical resistance of all materials varies with temperature. A coil of wire whose resistance is measured is a *resistance thermometer.* A *thermistor* is a resistance thermometer made of a bead of non-metallic material.

The radiation spectrum of a hot body varies with its temperature. The temperature can be measured remotely by comparing the appearance of the radiation with the appearance of a calibrated lamp filament heated by an adjustable voltage. The voltage is adjusted until the colors match. The temperature is then read from the calibration curve of voltage vs. temperature. This is a *radiation pyrometer.*

The maximum temperature reached in a furnace is recorded by *pyrometric cones*. These are small pyramids of ceramic calibrated to bend and droop until the tip touches the support surface at a particular temperature. A spectrum of different cones is put into the furnace. The highest rating to droop indicates the highest temperature reached.

7.8 SIGNAL TRANSMISSION

An on-off or analog electrical signal proportional to the measured parameter is sometimes converted to a modulated frequency or to digital form for transmission by wire, light beam, optical fiber, radio, or sound.

When there are too many parameters to be carried simultaneously by the available communication channels, for example in monitoring a complicated rocket during its launch, the parameters may be sampled and transmitted periodically at intervals corresponding to the maximum rates of change expected.

7.9 ENVIRONMENT

Transducers are used from ocean depths to outer space. They must withstand—and not pollute—environments as diverse as chemical plants and hospital operating rooms. The art is large, diverse, and growing.

CHAPTER 8

Controlled Motion

We now enter the dramatic realm of action: steering airplanes and missiles, orienting antennas and satellites, driving machine tools along contours, making integrated circuit masks, and doing most of it automatically or remotely.

8.1 ORIENTING

Orienting is controlling the angular position of a device.

8.1.1 Aiming

Aiming is orienting a device so that its axis is aligned with an object in the surrounding space; it does not require any linear displacement. A telescope is aimed.

The aimed device is usually mounted on a piggyback stack of two single-axis rotary joints. The stack may be based either on the ground or on a vehicle that itself is in motion. Sometimes a third rotary joint is added to the stack to prevent *gimbal lock,* which occurs when the aiming axis coincides with the axis of one of the rotary joints. *Pointing* is another word for aiming. A set of rotary joints is sometimes called a set of *gimbals,* particularly when it supports an instrument such as a gyroscope or compass.

Aimed devices include antennas, telescopes, searchlights, lasers, acoustic horns, cannons, rocket launchers, and hose nozzles.

8.1.2 Satellites

Satellites have no mechanical connection to a supporting base that can provide a resisting torque. Satellite orientation techniques include:

1. Spinning the entire satellite like a gyroscope rotor *(spin stabilizing)*

2. Firing pairs of small jets to generate torque, as in Hero's ancient steam engine. Large jets are small rocket engines and small jets use stored cold gas. In time, fuel or stored gas becomes exhausted, ending the useful life of the satellite.

3. Spinning small flywheels. Inertia wheels are small flywheels on servomotors. To apply a clockwise impulse to the satellite, a counterclockwise impulse is given to the flywheel; total impulse remains zero as pronounced by Newton. A full system requires three orthogonal inertia wheels. If there is a continuing external disturbing torque, the flywheels will gradually be accelerated to their speed limit so they must be slowed down and their momentum unloaded by one of the other means in this list. Inertia wheels provide a very fine adjustment of orientation.

4. Gravity gradient has been used to orient a number of satellites. Because gravity varies inversely with the square of the distance from the center of the earth, an elongated satellite tends to have its long axis torqued into the vertical. (The torque is very small, but the disturbing torques are still smaller, and a hundred feet of unrolled tubing makes a satellite very elongated.) Damping the pendulum-like oscillation is done with slightly shorter extension tubes pivoted in the horizontal position and damped with respect to the satellite body. (The moon is a slightly elongated satellite that oscillates through a small angle. The astronomers' word for the oscillation is *libration*.)

5. Electric currents in coils on the satellite react with the earth's magnetic field to generate orientation torque.

8.1.3 Guns and Missile Launchers

The paths of projectiles are curved by gravity and wind. Furthermore it takes time to reach their targets, which may be moving. Therefore, the launch-aiming axis is oriented in a computed direction such that the projectile trajectory will reach the target. (Guided missiles are steered after launch.) The gimbal and drive system is the same as if the gun or launcher were to be pointed directly at the target.

8.1.4 Examples Of Aimed Devices

1. *Antennas.* Antennas vary in size from a 2-inch wire on a cellular telephone to a 200-foot diameter parabolic dish. Large antennas to detect radiation from outer space are called *radio telescopes*. Antennas used in radar, communication with space vehicles, and as radio telescopes are aimed on servo driven rotary joints. (The world's largest dish, in Puerto Rico, is carved from a mountain valley and is moved only by the earth's motion, but its detector assembly is moved around its focus for fine aiming.)

Antennas used in fixed systems, such as microwave relays or household televisions, have adjustable rotary joint mounts but have no EM drives; they are clamped in the aimed position.

2. *Telescopes.* Telescopes are mounted on rotary joints, usually servo driven. The rotary joint systems of most astronomical telescopes have one axis parallel to the earth's axis, so the earth's rotation can be subtracted by a single motor. This is called an *equatorial mount,* Fig. 8-6. Some recent telescopes are mounted on a two-axis stack comprising azimuth and elevation rotary joints; these assemblies are much more compact than older telescopes, but they require synchronized motions of both axes. They were made possible by improved accuracy in synchronized motion control. *(Azimuth* is rotation about the vertical axis, that is, yaw, and *elevation* is rotation about the horizontal axis, that is, pitch.) The Orbiting Astronomical Observatory (Hubble telescope) is a satellite in orbit around the earth; its entire body is steered by EM inertia wheels, magnetic coils, and small rockets controlled by EM valves.

3. *Searchlights.* Now obsoleted by radar, large searchlights for antiaircraft use (and now for advertising use!) are mounted on motor-driven azimuth and elevation rotary joints. (The original advertising searchlights were war surplus, with carbon arc lamps, and with reflectors six feet in diameter.)

4. *Acoustic horns.* Also obsoleted by radar, large acoustic horns were coupled to human ears and used for detecting and locating enemy aircraft during World War I. They were mounted on azimuth and elevation rotary joints and driven either by hand or by motor.

5. *Ordnance.* Both ground-based and vehicle-based cannons and rocket launchers are mounted on servo-driven rotary joints. Their servos are automatically driven to subtract the motion of the supporting ship, tank, or airplane, and they are usually aimed by human, radar, or laser designator. (The use of a driven gimbal to compensate for a ship's roll was originally rejected by Navy officers who demonstrated that a team of men was not strong enough to swing a heavy cannon up and down at a ship's roll rate. Not having studied dynamics, they could not understand that the team need not swing the cannon at all. They must merely overcome bearing friction while the cannon orientation remains stationary and the ocean rolls the ship under the cannon.)

6. Human-aimed optical telescopes and range finders may be on hand-driven rotary joints. Transducer pickoffs tell a fire control computer the direction and range to the target. Machine guns and some small cannon and rocket launchers are on hand positioned rotary joints.

A designating laser is fixed to an aiming telescope and is directly moved by a human.

7. A rifle or pistol is mounted on that great complex of rotary joints, drives, sensors, and controller, the human body, and is aimed by human power.

8. Gyroscopes for stabilizing ship roll. Massive gyroscopes were experimentally used to exert roll torque on ships to maintain orientation of the ship on its roll axis. They were mounted on a gimbal permitting them to pitch but not to roll. As the sea applied roll torque on the ship, the gyros precessed on the pitch axis and resisted rotation in roll, thereby holding the ship steady in roll. The system worked but was too expensive. Present practice provides the ship with the equivalent of airplane ailerons. They generate roll torque by inclined plane reaction on the water as the ship moves forward. These roll ailerons are controlled by an *instrument* gyro.

9. A large class of aiming devices is machine tool rotary tables and heads. They do not launch missiles but they hold workpieces and machining heads. Their one-, two-, or three-axis orientations must be extremely accurate, stiff, and strong. For numerically controlled contour machining, these axes are synchronously driven from a program.

8.2 AIMING HARDWARE

8.2.1 Rotary Joints

There are four principal geometries for rotary joints: gimbal, universal, cantilever, and open axis.

1. *Gimbals.* See Fig. 8-1. A pair of gimbals comprise outer frame *(gimbal)* 1 carrying a pair of bearings 2a and 2b on axis 3. Inner frame (gimbal) 4 carries a pair of bearings 5a,5b on which are pivoted load 6 on axis 7. A limitation of gimbals is the occasional interference of the load or its optical or projectile path with the gimbal or bearing bodies. Therefore, other structures have evolved, as described below.

2. *Universal Joint.* (Figure 8-2 shows a *universal joint* (Hooke's Joint) for transmitting rotation between non-parallel shafts 9a,9b. The *U-joint* is a pair of rotary joints 6a,6b whose axes 7a,7b are at right angles to each other. They are pivoted to a common part, the central cross 8. Two universal joints with a shaft between them transmit rotation between input

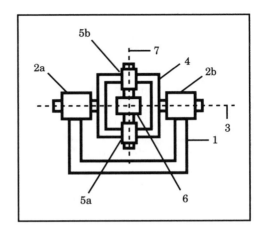

Figure 8-1   Gimbals

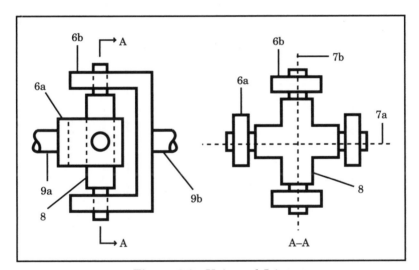

Figure 8-2   Universal Joint

and output shafts that are neither parallel nor coplanar. If the input and output shafts are parallel and if the universal joints have the same orientation, the transmitted motion is correct at all angles of rotation. (This assembly was invented by Robert Hooke, a contemporary and competitor of Isaac Newton. He also discovered *Hooke's Law* of proportionality between force and displacement for elastic bodies such as springs.)

3. *Cantilever Rotary Joint.* Figure 8-3 has its two bearings 11a,11b on the same side of device 12. It supports the device with a structure 13 extending outward from the bearings and base 15. There can be no getting in the way within a hemisphere on axis 14.

Figure 8-4 shows a single large-diameter ball bearing that is the equivalent of the two bearings in Fig. 8-3. Derricks, surveying transits, and machine tool rotary tables have cantilever rotary joints.

4. *Open-Axis Rotary Joint.* Figure 8-5 comprises an arc shaped slide 21, guided by a matching track 22, and rotating about axis 23. Device 24, such as a camera with its optical axis 25 passing through rotation axis 23, can rotate without displacing its optical axis. This ability is also true of other rotary joints, but mechanical considerations, such as overhung weights and interference of bodies, dictate which form of rotary joint should be used in each particular system.

Different devices require a *stack* of different numbers of rotary joints. For example, a shipboard torpedo launcher moves only in yaw since the

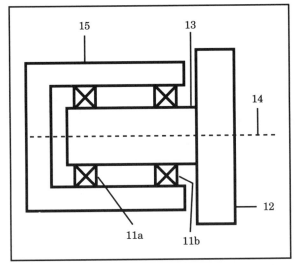

Figure 8-3   Cantilever Rotary Joint

Figure 8-4   Single-Bearing Rotary Joint

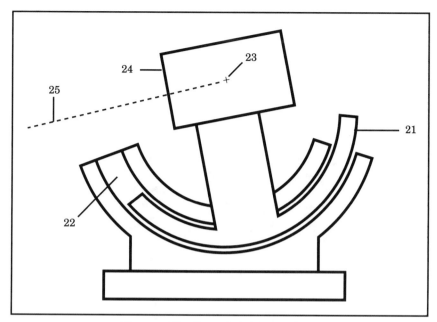

Figure 8-5   Open-Axis Rotary Joint

pitch of the torpedo during launch need not be controlled. A cannon is mounted on yaw *and* pitch joints but requires no roll orientation because its projectile need not be controlled in roll position. Some antennas require motion on all three axes: yaw, pitch, and roll.

A magnetic compass must remain level and must not turn in yaw relative to the vehicle, so it is supported by pitch and roll gimbals that are positioned by gravity. The needle is on a yaw bearing. The center of gravity of the compass is made well below the gimbal axes in order to provide gravity torque for pitch and roll orientation. The word *gimbal* was first used for magnetic compass mounts.

The usual stack sequence is a yaw-axis cantilever rotary joint starting with ground [Fig. 8-4]. It carries the pitch-axis joint, which is a gimbal. Some radar antennas also have a cantilever pitch axis.

For an aiming set, *ground* really is the earth for a land-based system, but it is the body of the vehicle for a vehicle-based system.

Large bearing diameters and hollow, notched shafts are sometimes used to circumvent obstruction of the device's working axes. The Palomar Mountain 200-inch telescope is a magnificent example, Fig. 8-6. Gimbal assembly 1 rotates about axis 2 which is parallel to the earth's spin axis. The lower gimbal bearing 3 is conventional. The upper gimbal bearing is a horseshoe-shaped member 4. Its outer cylindrical surface is the journal

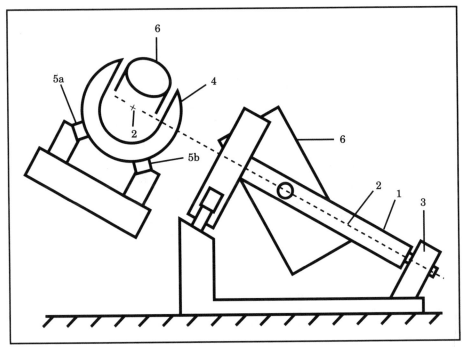

Figure 8-6   Telescope Gimbal

that rests on twin hydrostatic bearings 5a,5b [Section 2.8.3]. Thus telescope 6 can see along axis 2 without interference.

A common stacking sequence for an aimed device is ground (or vehicle) carrying a cantilever yaw rotary joint carrying a gimbal pitch rotary joint carrying the device. Such a stack for a camera is usually called a *pan-and-tilt* ("pan" being short for panoramic motion).

To compensate for the vehicle's orientation, a control system may require a coordinate transformation algorithm to convert a set of orientation parameters in one coordinate system to another coordinate system. For example, the orientation of a cannon mounted on a ship is most easily describable in the coordinate system of its axis stack on the ship. The ship itself has an orientation in pitch, roll, and yaw in a coordinate system based on the earth. Since it is necessary to orient the cannon with respect to the earth, it is necessary to do a coordinate transformation on its ship-based coordinates to get its earth-based coordinates.

8.2.2 Ball Joints

A ball and socket is a full three-axis rotary joint without stacking. Unfortunately, there is no known way to drive it mechanically except by torquing the carried device directly. It is used for an adjustment which is then clamped, such as a tripod camera mount, and for a coupling that transmits force but not torque, such as a spherical rod end.

(Actually, a ball can be rotated by three friction wheels, each having a rim carrying free cross rollers, but the system has very limited utility.)

Important uses for a ball joint are the following:

1. When suspending the stable platform of an inertial guidance system, the platform assembly is accurately balanced with its center of gravity at the center of the ball, air bearings [Section 2.8.3] are between the ball and the socket, and drive is by air jets exerting reaction torques on the stable platform assembly in just the same way that microrockets orient a satellite.

2. The reflecting mirror of the Sidewinder missile IR sensor is spun through a ball joint and acts as its own gyroscope to maintain angular orientation.

A stack of three single-axis rotary joints, each carrying the next one, is equivalent to a ball joint. For example, a yaw joint carries a pitch joint carries a roll joint. Sometimes a fourth joint tops the stack to prevent *gimbal lock* that occurs when the device axis coincides with one of the joint axes.

8.2.3 Drives

The drive for most rotary joints is a servomotor, a *drive train* of speed reducing gears, and a final worm and worm gear. Instead of a worm gear, a spur gear train may be used; usually, the final gear *(bull gear* or *ring gear)* being the largest permissible diameter. Sometimes the final stage has bevel gears instead of spur gears. In some very large machines that do not need high accuracy, the ring gear is replaced by a chain wrapped around a drum and engaged by a small-diameter drive sprocket.

Gear drives have backlash due to imperfect gear tooth shape, gear roundness, gear eccentricity, and gear axis positioning. One cure is to load the driving teeth always in the same direction with a spring-loaded reverse gear. Another is to provide a second motor and pinion which torque in the opposite direction from the drive pinion, also to load the drive teeth always in the same direction [Section 8.4.5].

For high performance (that is, fast and accurate) systems, there is gearless direct drive. Servomotors of large diameter and high torque are built into the rotary joint and drive it directly, without gears. Usually the stator is on the inside and the rotor is fixed to the rotating part. This system eliminates the inertia, backlash, and elasticity of gearing and has less net drive train inertia.

8.3 STEERING

8.3.1 General

Orienting a vehicle such as an airplane differs from orienting a ground-based device such as a telescope. The telescope is oriented by torques reacting on the ground or a supporting structure. The vehicle is oriented by torques reacting on environment to which it is not fixed (land, water, air, or vacuum). In the extreme case of a vehicle in vacuum, the torques can react only on their own exhaust. Vehicle torquing is called steering.

A vehicle may have one, or two, or three degrees of angular freedom. For example, a car or a ship is free to yaw but not to pitch or roll. (Yes, cars and ships have a limited responsiveness to pitch and roll torques, but they are restored to their equilibrium orientation by the resilience of their suspensions or by their stable buoyancy, not by active control.) Other vehicles in water, air, or space are actively controlled in yaw, pitch, and roll.

A moving vehicle is steered in air or water by rotatable *vanes* at one end of the vehicle. These are *rudders* on the yaw axis, *elevators* or *canards* or *diving planes* on the pitch axis, and *ailerons* on the roll axis. A vane is deflected and diverts the air or water stream to one side or the other; the reaction force on the vane urges its end of the vehicle to the opposite side.

Helicopters have a complex of vanes in their main and tail rotors. The main rotor vanes have variable angle of attack programs within each revolution and provide both orientation and propulsion.

Some ships have auxiliary bow propellers thrusting sideways to help steer in close spaces. Some ships have ailerons to reduce roll.

Rockets above the atmosphere cannot steer with vanes because there is no air for the vane to react against. They steer by tilting their engines, mounted on rotary joints, to produce a transverse component of thrust. The German V2 used steering vanes in the rocket exhaust. These tilted the rocket engine exhaust without having to tilt the rocket engine.

On the ground, auto wheels generate steering torque by ground friction, caterpillar tractors and tanks use differential ground speed of their two tracks, and railroad trains use tracks guiding their wheels for steering torque.

8.3.2 Propulsion

Most vehicles are propelled along their principal axis. Examples are airplanes, cars, tanks, ships, rockets, and torpedoes. (Helicopters are more complicated.) Steering such a vehicle is by controlling its orientation. Even gravity-propelled smart bombs and gliders are steered to control their path.

In some cases, the axis of the propulsion device is inclined to add a transverse component of force for steering. Examples are an outboard engine on a boat, a water nozzle on a jet ski, and the rocket on a missile.

8.4 POSITIONING

8.4.1 Kinematics

Kinematics is the geometry of motion. *Positioning* as used here means controlling all six axes of a body, three linear (X, Y, Z or E-W, N-S, up-down) and three angular (Roll, Pitch, Yaw).

A single slide on a machine tool has one degree of linear freedom.

A car or a ship moving on the surface of the earth has two degrees of linear freedom and one degree of angular freedom.

An airplane or a submarine has all three degrees of linear freedom and all three degrees of angular freedom.

A stack of three orthagonal linear slides and three rotary joints provides all three degrees of linear freedom and all three degrees of angular freedom. Many machine tools have such a stack, and some have redundant slides and rotary joints for other reasons.

A steered vehicle can displace only along its propulsion axis. Therefore, it is necessary to change orientation first in order to change position. Conversely, in a stack of slides and rotary tables in a machine tool, any degree of freedom may be exercised without changing any other degree of freedom.

Most applications of *steering* are intended to achieve final *positioning*. With missiles, launch aiming is also intended to achieve final positioning at a target.

So far we have dealt only with aiming, which is orientation in a fixed position, and with steering, which is orientation of a self-propelled vehicle toward a target position. We now consider means to displace a body to a desired position without a propulsion system in the body itself.

8.4.2 Machine Tools

For the purpose of this study, we will include as *machine tools* not only the cutting machines usually called machine tools but also semiconductor mask-making machines, some laboratory instruments, and a variety of other positioning machines, including robots. All, except some robots, use geometrically similar positioning and orienting systems for a body that may be a workpiece, a tool, or a measuring instrument.

Linear displacements vary from a small fraction of an inch to many feet; rotations may be up to 360°. Linear accuracy errors permitted vary from an inch to a fraction of a light wave; angular accuracy errors may be down to microradians.

A body may have one, two, or three degrees of freedom in position as well as one, two, or three degrees of freedom in orientation. The positioning geometry for most machine tools is a stack of single-axis motion elements: linear slides and rotary joints. Rotary joint construction has already been discussed; we now consider linear slides.

8.4.3 Linear Slides

A linear slide comprises a base, straight line bearings on the base, a table that moves in a straight line along the bearings, drive means, and transducer means. The base may be mounted on the table of another linear slide or on a rotary joint, and another linear slide or a rotary joint may be mounted on its table to form a stack or part of a stack. In microscopes, the stack is called a *stage*. Other words used are *deck* and *platform*.

For example, in knee-type milling machines, a vertically moving *knee* carries a horizontally moving *cross slide* that carries an orthogonal horizontally moving *table* that may carry a *rotary table* or a *dividing head*. A dividing head may have one or two orthogonal axes of rotation. Some machine tools have a motion stack for the workpiece and a separate motion stack for the tool.

8.4.4 Materials

Most machine tool bases and tables are made from iron alloy or aluminum alloy. The metal may be cast, welded, or machined from solid.

Other materials are also used:

1. *Granite* is rigid, stable, and can be machined to close tolerances. If dented, it chips but does not form burrs, and close clearances are not interfered with. This permits the use of air bearings integral with the granite blocks. Granite has high damping of vibration.
2. *Filled polymers* have low specific gravity but otherwise resemble granite.
3. *Reinforced concrete* has been used for very large machine tools. Metal inserts were provided for slides and drives.
4. *Honeycomb* is used for some laboratory tables.

8.4.5 Linear Drives

EM and direct engine drives for vehicle positioning on the ground use wheels, caterpillar tracks, cable or chain, cogwheel and rack, or linear motor. For example, cable is used in elevators, San Francisco cable cars, some Swiss mountainside railroads, ski lifts, and Disneyland boats. Cogwheel and rack is used in other Swiss mountainside railroads and a variety of less romantic machinery where it is called rack and pinion.

Drive for vehicle positioning on or under water is by propeller, water jet, or water wheel.

Drive for vehicle positioning in the air is by propeller or jet. Drive for rocket positioning in the air or in space is by rocket jet.

Linear slides in machines are driven by:

1. Screws with solid nuts, usually with acme threads for minimum friction, lubricated

2. Ball nut screws, which are analogous to recirculating ball bearings. The balls roll in helical grooves in the screw [Fig. 2-10].

3. Rack and pinion. Either the rack is on the base, extending the full length of travel, and the pinion is on the moving table, or the positions are reversed.

4. Sprocket and chain. A pair of drive sprockets is located at the ends of travel. They form the chain into a closed loop. One point of the chain is fastened to the table. This scheme is cheap but is inaccurate. The chain droops and stretches. Frequency response is low because of chain elasticity.

5. Friction rollers, as in a roller conveyor

6. Hydraulic cylinder with EM control valve

7. Linear electric motor built into the base and table

8. A linear drive using a helical friction drive roll is shown on Fig. 8.7. Payload 1 carries control frame 2 on which is driven roll 3. At least part of the weight of payload 1 is carried by driven roll 3, resting on long driving roll 4. Driving roll 4 is continually rotated in an unchanging direction by a motor, not shown. Control frame 2 is turned on axis 5 by an EM control mechanism, not shown. *The control frame angle establishes the angle of helix 6 of the path taken by driven roller 3 on driving roller 4.* Wheel 3 moves in both directions, although the rotations of wheel 3 and roller 4 are always in the same direction. Very smooth accelerations and decelerations are obtainable from simple EM controls. This drive has been used on assembly machine pallets, on large material-handling conveyors, and on tiny phonograph cartridge carriers.

An example of extreme accuracy in a steering servo is the steering of the sensing laser along the record track of a compact disc.

8.4.6 Backlash

Backlash is present in most rotary and linear drives. It can be ignored if final positioning can always be made in the same direction. Otherwise, it is removed by elastic loading of the drive, typically by double nuts with

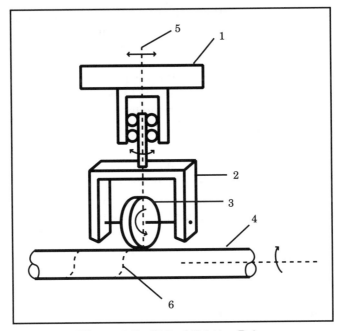

Figure 8-7   Helical Friction Drive

a spring between them, or by providing two motors torquing in opposite directions.

8.4.7 Transducers

Feedback transducers for both rotary joints and linear slides are described in Chapter 7.

8.5 CONTROLS

8.5.1 Open-Loop Controls

An *open loop* control moves the part in accordance with a preestablished *program* without using feedback information about the actual position of the part. Examples of open loop controls are:

1. A train rolling on a track
2. A package transported by a series of conveyors
3. A machine tool slide driven by a cam
4. A motion limited by a hard mechanical stop
5. A factory transfer line
6. A two-position device driven back and forth by a fluid power cylinder

7. Fixed fins that orient a dart, arrow, or gravity bomb along its ballistic trajectory

8. A manual control is often described as open loop but it really is a closed loop with a human being as part of the loop.

8.5.2 Closed-Loop (Servomechanism)

Aiming, steering, and positioning are usually done by feedback control systems—*servomechanisms* or *servos* for short. Automatic feedback control replaces the person's eyes with transducers, brain with amplifiers, and hands with actuators. This is not science fiction robotics. Only very limited and precisely defined human functions can be replaced by machines, but those machines may be faster, stronger, and more accurate than people [13.1, A23].

Some examples:

1. The oldest feedback controller is a person, as when a sharpshooter aims a rifle or a driver steers a car. The person senses with eyes an error between the actual and the desired direction, decides with brain what to do, and corrects the error with hands.

2. Servo aiming is usually based on sensing radiation from a target: light, heat, sound (including sonar), radio, or radar. The radiation can produce no direct mechanical force and so is used as a reference signal to the transducers of closed-loop servos.

Continuous aiming at a moving target is called *tracking*.

3. *Guidance* of a moving vehicle such as a boat, torpedo, airplane, or missile. The guidance system may, for example, follow a radio beam or sense the earth's magnetic field or be inertial, sensing angular orientation with gyroscopes and linear motion with accelerometers.

4. *Web guidance.* A *web* is a continuous ribbon of material passing through a machine. The web may be steel, cloth, or paper. The edge is sensed either photoelectrically or pneumatically, and the web is steered laterally to the specified position by tilting certain of its support rolls.

5. *The industrial automatically guided vehicle (AGV).* Wires are embedded in the floor of a factory aisle. AC produces a magnetic field along the aisle. Sensing coils in the AGV send feedback to the servo that steers the AGV along the embedded wires.

6. Items 1 through 5 are position servos. The cruise control in your car is a rate servo; it controls the speed of your car.

Chemical industries use many servos, usually controlling a fluid with a valve, to maintain temperature, mixture ratios, and many other parameters.

Closed-loop systems contain error-sensing transducers, of which there are very many [Chapter 7]. Angular error sensors include gyroscopes, radars, and directional IR detectors. Linear (that is, displacement) error sensors include radars, radio and optical interferometers, differential transformers, and EM encoders. Closed-loop systems have amplifiers, stabilization circuits, and error correctors such as the servomotors in Chapter 3 and the actuators in Chapter 6.

8.5.3 Telemetry and Remote Control

Wire telegraphy in the mid-nineteenth century and wireless telegraphy in the early twentieth century were the great leaps forward in the fast transmission of data. The addition of transducers and analog-to-digital converters (or direct digital encoders) made possible telegraphy of data without human intervention; this is *telemetry*. Video images as well as numbers may also be telemetered, usually by first digitizing the image.

Multiplex telegraphy, sending more than one message at a time over the same pair of wires, was first invented by Edison in the nineteenth century. Now a communication channel is divided into many subchannels of different bandwidths and time divisions and transmits the outputs of many transducers at the same time.

Satellites (and missile launches) telemeter many kinds of data after first digitizing them. Satellite data are first stored on magnetic tape in the satellite until a convenient time for radio transmission. The data are then stored on magnetic tape on the earth until a convenient time for analyzing it. (Early military surveillance satellites made photographs on film and the film was recovered from orbit by retrorocket, parachute, and catch planes.)

Many scientific and engineering experiments generate large amounts of data that are similarly digitized and stored in what might be called zero-distance telemetry. A special case is astronomy. This data are mostly spectrum analysis of electro-magnetic radiation from many sources (stars, galaxies, quasars, dark matter, etc.). The "laboratory" is a giant telescope on a mountaintop far from the astronomer's university. It is expensive and time consuming to get to and uncomfortable to be at. Now it is possible for the astronomer to command the telescope by remote control, to see the data in real time on an oscilloscope, and to record it on tape for later analysis.

Remote control is command telegraphy, the obverse of information telegraphy. The two are commonly used together: a human reads the telemetry, makes a decision, and issues commands via remote control. If no human judgment and decision is needed, there is no need for the telemetry link except for record and review; the feedback and the control response can be local and automatic [13.2].

8.5.4 Numerical Control

Machine tools may be programmed to go to a series of discrete positions at each of which is performed an operation, such as drilling a hole. If the program takes the form of a series of dimensions, the process is called point-to-point numerical control (NC).

Machine tools may also be programmed to move in a continuous curved path along which an operation, such as milling, is performed. If the path is defined digitally, the process is called continuous-path numerical control.

Since the early 1950s, more and more machine tools have been made with numerical control. They have digital input, servomotor output to slides and rotary joints and analog-to-digital feedback transducers, *encoders*. Machine tools may have five or even more motion axes, driven synchronously, to cut complex three-dimensional shapes. Computer programs with human input from dimensioned drawings prepare the data; some programs prepare the data directly from CAD programs.

The control circuitry is essentially that of a digital computer.

(This art is now so highly developed that the most modern astronomical telescopes use it to drive azimuth and elevation motions simultaneously to compensate for the earth's rotation, instead of using the traditional equatorial drive.)

8.5.5 Program Memory

8.5.5.1 Analog Memory

A set of cams mechanically moving the slides and rotary joints of a machine comprise an analog open loop memory with high speed, great reliability, and close accuracy. However, the cost of designing and making the cams may be substantial, and making a change may be the same as starting over with a new set of cams. Large machine tools would require cams of impractical size. The most common cam-operated machine tool is the small automatic lathe, for example, the *Swiss automatic screw machine*.

For simple rectangular motions, a set of limit switches operated by cams on the slides and rotary joints are analog memories which set dimensions. A drum controller or similar sequence switch is sufficient and is commonly used. Such a system is closed loop with the limit switches as binary feedback transducers.

A common analog feedback control system uses a template or model of the finished part as a program memory. A transducer senses the difference in position between the template or model and the machine slides. The transducer may be electrical, such as an LVDT, or an hydraulic valve that directly feeds the slide cylinders. Molding dies that must reproduce a shape represented by a model are made in such machines.

8.5.5.2 Digital Memory

The first mode of digital data storage and entry was punched tape. The first use of punched tape was all mechanical, in the eighteenth century, in Jacquard looms. Each hole controlled a single crossover of threads in the fabric; thus very intricate patterns could be, and still are, woven.

Monsieur Baudot adapted the technique to telegraphy, with one cross row of five holes per character, whence we have Baudot coding and the Baud as a unit. Five holes were increased to six, then seven, then the present eight to accommodate more characters and error-checking schemes.

Punched tape was originally chained cards (Jacquard), then paper, then fiber, then Mylar as the demands for reuse increased.

Holes were originally sensed by spring-loaded mechanical pins that tried to enter the holes, then by air currents cooling thermal transducers if air passed through the holes, and then by photoelectric cells sensing light that passed through the holes. Holes were sensed one character at a time and then in multicharacter blocks. Making punches and readers for punched tape was once a substantial industry. Punched-tape technology was similar to the punched-card technology used in computers.

The next step, of course, was the replacement of punched tape with magnetic tape that could be read faster, was more compact, was less subject to error in reading, and could be reused more times.

The next step was one of data reduction: CNC (computer numerical control) machines did their own computation of curve points from the equations and parameters defining the curves. Among other benefits, programming is easier on the shop floor and it is also possible to program a machine tool over wires from a central control office (Direct Numerical Control, DNC).

Meanwhile, magnetic tapes have been replaced by magnetic discs, except for backup.

Some numerically controlled robots can be programmed in what is called "teach mode." Manual switches or transducers on the robot joints command motions; electronic memory (magnetic or solid state) records and later plays back the program. A practical use is programming a paint spray gun path that is not mathematically definable but that is the actual path used by a skilled painter [13.1].

An unusual digital memory is the "shish kebab" used in the U.S. Postal Service letter-sorting machine, Fig. 8-8. One memory accompanies each letter passing through the machine. Shaft 1 carries a plurality of

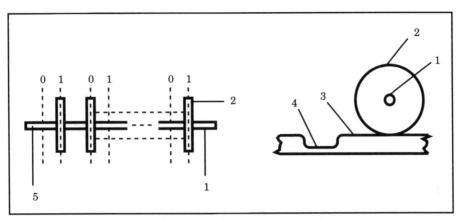

Figure 8-8   Letter-Sorting Machine Memory

wheels 2 rolling along path 3. Each wheel is spring detented into either its 0 position or its 1 position. At each sorting position, the path has a cross row of depressions 4 at either the 0 or the 1 position of each wheel. If every depression is in line with its corresponding wheel, the entire shaft dips the depth of the depressions; but if even one wheel does not cross a depression, it supports the shaft. When a shaft dips, it releases its letter, which drops into a sorting box. (The device is a mechanical digital memory with a parallel mechanical reader. Admittedly, it is not EM.) Millions of these memories have been in service for many years. The inventing process and a photograph of the machine are shown in [A3, pp. 19–22].

CHAPTER **9**

Contact Makers
and Electrodes

9.1 CONTACT MAKERS AND ELECTRODES

Electrical contact-making devices include the following:

- Relays and contactors
- Circuit breakers and fuses
- Motor starters
- Push buttons
- Limit switches
- Selector switches
- Telephone dials
- Ignition breakers, distributors, and spark plugs
- Buzzers
- Connectors
- Trolleys
- Slip rings
- Electrodes

Most of these close and open circuits by touching and separating—*make* and *break*—for example, relays and circuit breakers. Others slide without separating, for example, commutators and slip rings. Others provide spark gaps. Others conduct to things not a part of the device, for example, welding, chemical, and medical electrodes.

172

The materials for contacts and electrodes were described in Section 1.1.4. (This would be a good time to review this section.) More details on contact and electrode materials will be given in Section 9.13.

9.2 RELAYS AND CONTACTORS

9.2.1 General

Relays and contactors are driven by AC or DC electro-magnets, and they open and close contacts. For example, Fig. 9-1 shows a relay comprising electro-magnet 1 which pulls armature 2. Mechanical linkage 3 transmits armature motion to one or more contacts which may include a *normally open* contact 4, a *normally closed* contact 5, and a *transfer* contact 6.

Relays and contactors differ only in size; contactors are large relays. Relays are usually thought of as performing logic, while contactors are thought of as connecting power to loads, but these are not rigorous definitions; there are many *power relays* which switch load power, and many contactors have low current contacts in control circuits as well as high current contacts in power circuits.

9.2.2 Uses for Relays

Telephone switching and industrial control logic was done entirely by electro-magnetic relays until the advent of solid-state logic. Much industrial control logic is still done by EM relays.

Relays provide complete isolation of their controlled circuits.

Solid-state logic first appeared in discrete logic packages and then in microprocessors. There is a special class of industrial microprocessor called a programmable logic controller (PLC or PC). The PLC is pro-

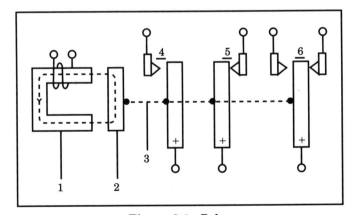

Figure 9-1 Relay

grammed with the same kind of logic and the same kind of *relay ladder logic diagram* as are relays. Figure 9-2 shows the notation. In the top *rung of the ladder*, when push button A is pressed, it energizes the coil of relay B. When B operates, its contact energizes coil C provided contact D is not open, and also lamp F if contact E is closed. Each rung of the ladder is a statement in Boolean algebra. For a complicated machine, the ladder may have hundreds of rungs, some rungs may be branched as in the lower rung of the figure, and there may be many elements in series in a single rung.

Modern telephone switching is entirely digital and solid state [A22], but for a while, reed switch relays were used (the reed switch was developed for telephone switching). Before then, crossbar switches were used, and before then, open contact relays and selector switches were used. *(Telephone-type relays* are still made, but not for telephones.)

An early electrical digital computer used relays. Computer relays were displaced by vacuum tubes which were displaced by separate transistors which were displaced by integrated circuits. Some day, perhaps, optical devices.

Power switching is sometimes done by electro-mechanical contactors and sometimes by solid-state transistors or silicon controlled rectifiers *(SCRs).*

The choice between electro-mechanical and solid state is sometimes made by psychological inertia and stubbornness (euphemistically called *conservatism)* either by engineers, or by managers, or in recognition of resistance to innovation by working-level technicians. Choice on the basis of economy, reliability, or other non-emotional factors of merit is sometimes secondary.

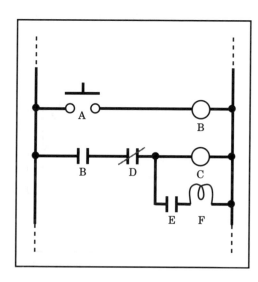

Figure 9-2 Relay Ladder Logic
Diagram

9.2.3 Construction

Relays and contactors are packaged in a variety of enclosures, from none at all through dust tight to vacuum tight.

Most operating electro-magnets are non-polarized and pull against a restoring force from a spring or from gravity. Some *sensitive* relays are driven by polarized electro-magnets with either spring return or reverse-current return.

Some contactors are driven by air cylinders. A small air cylinder can do a lot more work than a big electro-magnet, but it requires an air supply and an electro-pneumatic valve or an air logic valve to control it.

Although polarized relays usually have only a single contact and are intended to operate on extremely low power, many relays and contactors have several contacts; some *gang relays* have as many as 100. The same relay or contactor may have contacts of widely different ratings, some for power circuits and some for control circuits.

The expected life of a relay or contactor is many millions of operations. Many motor-starting contactors have easily replaced contacts to take advantage of the long life of the rest of the device.

9.2.4 Contact Configurations

Relays and contactors have a variety of basic contact arrangements. Multiples of these arrangements are combined into individual devices. The word *normal* is used to mean the condition with the coil deenergized, even if, in the circuit used, the coil is usually energized. See Fig. 9-1.

- Normally open = *Form A*
- Normally closed = *Form B*
- Transfer = *Form C*
- Make before break = *Form D*

One construction of a contact pair, Fig. 9-3(a), comprises a fixed contact 1 and a moving contact 2 with the moving contact connected to the circuit by either a cantilever spring on which it is mounted or by a flexible *pigtail* 3. (In a big contactor, this "pigtail" may be a massive set of flexible copper braids.)

Another construction of a contact pair, Fig. 9-2(b), comprises two fixed contacts 4,5 and a moving pair of matching contacts 6 which bridges the fixed contacts; the configuration is called *double break*. Flexible pigtails are subject to fatigue breakage, so double break contacts are preferred when long life is needed.

Some relays use reed switches, others use mercury switches, others use snap-acting *microswitches*. (Micro-Switch is a trademark of Honeywell, Inc., but generic usage is common.)

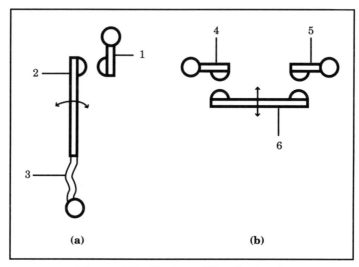

Figure 9-3   Single and Double Break

The contacts in industrial control relays and contactors may gradually burn away by arcing, but in many designs they can be replaced easily without replacing the entire relay or contactor.

9.2.5 Actuators

Relays and contactors are actuated by the following devices, already described in Chapter 6:

1. DC square law electro-magnets
2. DC polarized electro-magnets
3. AC electro-magnets
4. d'Arsonval meter movements with a contact on the pointer
5. Watt-hour meter movements with selected AC voltages or currents in the two driving coils
6. Air cylinders or hydraulic cylinders (usually only for large contactors and circuit breakers)

Relays and contactors which control reversing motors are often mounted in pairs with a mechanical "seesaw" interlock lever between them so an electrical fault which energizes both at the same time cannot cause a short circuit. Others have mechanical latches so that a human action is required either to release or to operate the relay.

9.2.6 Analog Sensing Relays

Most relays are binary digital devices: The input is on or off. They were the original components of automatic digital computers after Babbage's mechanical computer and are still used by the millions in industrial control systems. In the long run, most usages will probably be replaced by solid-state devices and systems, such as programmable logic controllers, but for a variety of real-world reasons, mostly human inertia, that time is not yet here.

In addition to digital logic and power switching, some relays are used to measure analog parameters and operate or release in accordance with those parameters. For example:

1. Time delay
    a) DC relays. These may have built-in electro-magnetic time delays, either between energizing the coil and operation of contacts *(make-time delay)* or between deenergizing the coil and release of contacts *(break-time delay)*. A copper ring around the core near the armature end produces a make-time delay, and a copper ring around the core away from the armature produces a break-time delay. Both are for fractions of a second, but useful in sequencing relay operations.

    b) Thermal timers. Relay operating voltage heats a bimetal strip which bends and closes or opens a contact. Delays of several seconds are available.

    c) Pneumatic timers. A solenoid or spring presses a diaphragm which moves air through an orifice. At a particular displacement, the diaphragm operates a contact. These timers are adjustable by adjusting the orifice. Delays up to a minute are available.

2. Marginal operation. DC relays can be adjusted to operate above a particular current and not operate below that current. Similarly, they can be adjusted to release below a particular current and not release above it. A special case is a relay comprising a d'Arsonval meter movement with a contact on its pointer mating with fixed contacts at the ends of its travel.

9.2.7 Watt-Hour Meter Relays

The relay is a watt-hour meter motor with different coil connections. Displacement of the rotor operates a contact. These relays can measure time, voltage, current, power, impedance, phase unbalance, power factor, and so forth. They are the basic devices in power system protection. They have wide range, high accuracy, and high reliability even after a long period of inactivity.

9.2.8 Crossbar Switch

A crossbar switch is a remarkable kind of relay developed for telephone switching. Assume X, Y, Z orthogonal axes. The crossbar switch has a first array of ten conductors oriented in the X direction, a second array of ten conductors oriented in the Y direction and offset from the X array by a small distance in Z. There are ten electro-magnets each associated with a conductor of the X array and ten electro-magnets each associated with a conductor of the Y array, so there are $10 \times 10 = 100$ possible outputs. When one X and one Y electro-magnet are energized, they displace their associated conductors toward each other in the Z direction until they contact and close a circuit. There are actually pairs of conductors, not single conductors, in each array, four levels in all.

Crossbar switches were next to the last of the EM telephone switching techniques. Then came reed relays and then digitizing of voice signals and all electronic transmission. A modern digital system for connecting calls is still called a *switch*.

9.2.9 Telephone Dial

The old-fashioned telephone dial illustrates how reliable a complex EM device can be. Millions were installed in homes and offices, received no routine maintenance, and almost never failed.

We will assume that you are familiar with its externals and its use. The dial has a cam-operated contact, a speed governor, and a pulse counter. The pulse counter is the external ring of holes, the user's finger, and the finger stop. (Open loop and no mechanism or circuitry!) The speed governor is a centrifugal brake; above the preset speed, a brake shoe flies out and drags on a fixed brake drum. (No lubrication.) The cam-operated contacts are completely conventional. (Elegant design, now obsolete.)

9.2.10 Insulation

Relay and contactor contacts are used at all voltages, from microvolts to kilovolts. Dielectric failure may occur both from breakdown through air and from surface creepage. It is customary to specify both air and surface distances and to apply a high-voltage *(hipot)* test, usually twice working voltage plus 1000.

9.3 CIRCUIT BREAKERS

A circuit breaker is a contactor which reopens automatically if there is an overload. All but the largest are manually closed against a spring opening force, are mechanically latched, and are unlatched by an overload sensor and actuator. An overload may be a short circuit with current hundreds of times more than normal, so the interrupting capacity must be correspondingly high.

The overload sensor is either EM or thermal. It may have a time delay so it is insensitive to harmless transient overloads.

Utility-size circuit breakers are closed by air or hydraulic cylinders. They may be automatically reclosed several times so that transient faults like lightning strikes do not cause more than very brief power interruptions.

9.4 ARC QUENCHING

9.4.1 Arcs

When contacts try to interrupt a current to an inductive load, the decreasing current and its magnetic flux induces a voltage which tries to keep the current flowing. Furthermore, in a multikilovolt circuit, the circuit voltage also tries to keep the current flowing despite the opening contacts. These voltages ionize the air between the separating contacts, and an arc occurs. The arc can continue to burn and prevent circuit interruption, and it burns away the contacts. The following are techniques to extinguish the arc.

9.4.2 Magnetic Field

A magnetic field is provided transverse to the arc. The arc is driven across the field by electro-magnetic force just as if it were a current in a wire, its length is extended, and the greater surface area cools it. The process is called *magnetic blowout*. Actually, there always is some magnetic blowout because as the arc extends, it increases the flux linkage of the circuit.

For AC circuits, the magnetic field is generated by a coil in series with the contacts so the magnetic field is always in phase with the arc current. A similar coil or permanent magnets may be used for DC circuits.

It is customary to orient a circuit breaker so that the heat of the arc causes a convection current in the air which furthers the arc extension.

9.4.3 Multiple Break

The circuit is interrupted at two or more places, such as both sides of the power line. The arc generating voltage is divided and there are two or more sets of contacts to absorb the initial arcing heat. This divide-and-conquer cooling technique is extended by driving the arc into a stack of insulated metal plates.

In Fig. 9-4, current enters and leaves via conductors 1,2 and passes through contacts 3,4. When contact 4 swings away from contact 3, the resulting arc transfers from contact 4 to arc runner 5, and the arc 7 rises along arc runners 5,6 toward insulated metal plates 8. The arc divides into short arcs, each bridging the space from plate to plate, and each short arc is cooled by contact with a pair of plate surfaces.

9.4.4 Long Contact Gap

The contacts are spread wide to lengthen the arc and expose it to more cooling air.

9.4.5 ZigZag Path and Ablation

A zigzag path is provided into which the arc is blown magnetically or by a blast of air or other gas, thus increasing the arc length and providing cool surfaces alongside the arc to cool it, Fig. 9-5. The cooling surfaces may be either refractory and cool only by conduction, or they may be organic and give off gases *(ablate)* which further cool the arc by absorbing heat in the ablation process. (Ablation is also used to cool surfaces of bodies returning from space at very high speed.)

In a large *fuse,* the conductor is embedded in sand that cools the arc.

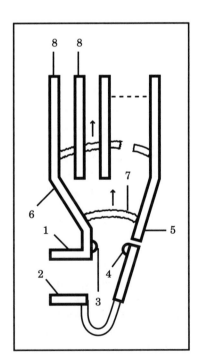

Figure 9-4   Multiple-Plate Air
Circuit Breaker

9.4.6 Cross Blast

Very large circuit breakers blow either compressed air or other gases transversely across the arc to extend and cool it. In Fig. 9-6 the contacts 1,1' are immersed in oil 2 in *explosion pot* 4. Oil decomposition gases 3 are produced by the heat of the arc and blow oil and gas through the arc and vents 5.

9.4.7 Pressurized Atmosphere Or Vacuum

Easily ionized air is replaced with a pressurized atmosphere of either hydrogen or sulfur hexafluoride, both of which have much higher heat absorption and higher ionizing voltages than air. Alternatively, the contacts are in high vacuum, which does not ionize at all.

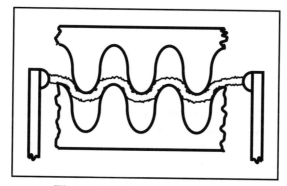

Figure 9-5   Zigzag and Ablation

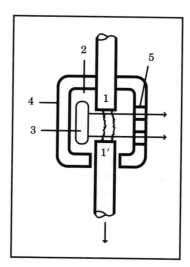

Figure 9-6   Oil Circuit Breaker

9.4.8 Opening at Current Zero

Since AC circuits have a current zero 120 times per second, regardless of faults, the control circuit delays contact separation until the next current zero. In utility systems, instrumentation relays select both the particular breaker to open for a particular fault and the particular instant to open it. These instrumentation relays may be based on the watt-hour meter movement described in Section 6.4.

9.4.9 Protective Networks

DC circuits can provide a network which bypasses the decaying current around the contacts. A resistance-capacitance (RC) branch in parallel with either the contacts or the load does this, as does a single diode across the load.

9.5 TRIPPING

A *tripping device* responds to an overload current and, usually, waits for a short time to avoid tripping on a transient. The time delay approximates the integral of i^2t (a measure of load heating) to permit the maximum delay consistent with protection of the load from overheating.

The tripping device must permit manual tripping independent of overload tripping, and vice versa. It must be *trip free;* that is, it must prevent manual reclosing from blocking the action of the tripping device in the continued presence of the overload. Mechanical linkages couple the tripping device, the manual operating lever or button, and the contact mechanism and latch. Additional linkage may couple these to a door lock. The overall mechanism may be much more complex than the circuit breaker itself.

The most common tripping device is an electric heater in series with the load and a mechanical element responsive to temperature rise and coupled to the circuit breaker latch. The mechanical element may be a piece of bimetal or it may be a solder pot with a spring-loaded rotor which turns when the solder melts.

Another tripping device is an electro-magnet with two armatures. The first armature is sealed in a container with viscous fluid. On overload, the first armature moves slowly into a position in which the magnetic circuit reluctance is sufficiently low to cause the second armature to move and trip the breaker. On dead short circuit, the current is sufficient to attract the second armature instantaneously.

Large breakers controlled by protective instrumentation *(relaying)* systems use electro-magnet trips switched on by the relays.

In distribution systems with echelons of circuit breakers in feeders and branches, relaying is provided to determine where the fault is and to open only those breakers necessary to isolate the fault, thus minimizing the extent of the power outage. Furthermore, since some faults are transient, such as from some lightning strikes, there is automatic reclosing and retrial to restore power as soon as possible.

A large circuit breaker under short circuit is part of a single-turn coil carrying tens or hundreds of thousands of amperes. Such a coil has a large expanding force because as it expands, its flux linkage increases. Therefore, the latching mechanism must be strong and rigid enough to not be broken or deformed or jammed by that force so that the relatively weak force of the tripping mechanism can release it.

A circuit breaker may be equipped with an additional ground fault sensing relay which trips the breaker instantly on sensing even a small ground fault current. Many small domestic breakers are now so equipped.

Solid-state switches are continually being increased in capacity for use in utility power systems [A49]. They will gradually supersede larger and larger contactors and circuit breakers, but it may be some time before they can interrupt fault currents between 100,000 and 1,000,000 amperes in a circuit driven by several hundred thousand volts.

9.6 MOTOR STARTERS

A motor starter is a combination contactor and circuit breaker. It is used to switch a motor on and off and to protect it against overloads.

A motor starter is actuated and released either by hand or by electro-magnet, and it latches either mechanically or electrically. (*Electrical latching* is done by closing a contact in parallel with the momentary contact which energized the electro-magnet in the first place.) A motor starter is also released by an overload tripping sensor. The sensor has a time constant which approximately matches the thermal time constant of the controlled motor. Thus the motor is not disconnected by its starting transient or by momentary overloads but only if it overheats.

If the motor is reversible, two contactors are used, one for each direction of rotation. If both were accidentally operated at the same time, there would be a short circuit so a mechanical interlock is provided to prevent simultaneous operation.

(This is one of the places where mechanism is considered more reliable than electrical logic. Electrical logic is subject to spontaneous or imposed defects in components, wiring, and power supply, but it can be far more sophisticated than mechanism for a smaller budget. Mechanism is subject to imposed jams from severe physical abuse. Both are subject to corrosion. Choice, or mix, is based on experience, judgment, and the specifics of the situation. Engineering is simpler if one is merely dogmatic and doctrinaire, but the result may be much worse.)

AC motors may be braked electrically by applying reverse voltage for a short time; this is called *plugging*. A *plugging switch* is a transducer on the motor shaft which senses the beginning of reverse rotation and commands stopping reverse voltage.

If AC line voltage falls too far the resulting current *increase* may burn out the motor so it is common to provide an undervoltage sensing relay which trips the starter.

9.7 SWITCHES

9.7.1 Two-Position Manual Switches

There are many manually operated contact makers. Examples are:

1. A push button on your telephone or computer keyboard

2. A multikilovolt utility system switch operated with an insulating pole

3. A low-voltage, thousand ampere knife switch in an electro-chemical plant. (To reduce arcing, two-position switches which must interrupt large currents are provided with spring-actuated portions to do the final snap-action interruptions.)

Push buttons may have any of a large variety of human interfaces, including color-coded buttons, large palm buttons, and internally illuminated faces which may also serve as indicator lights. A row of push buttons may be mechanically interlocked so that the last button pressed remains down and releases the button previously pressed.

Some computer keyboards are now using diaphragm switches. Figure 9-7 shows a laminated assembly of a thin, flexible plastic sheet 1, a center

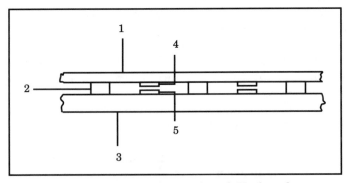

Figure 9-7   Diaphragm Switch Keyboard

plastic sheet with windows 2, and a rigid plastic sheet 3. Etched circuitry 4,5 is on the inner surfaces of sheets 1 and 3. The face of sheet 1 is printed to identify the windows. When a window area of sheet 1 is pressed its circuit metal 4 contacts the circuit metal 5 on sheet 3. Such panels are cheap, liquid tight, and easily cleaned but lack the positive feel of a conventional computer keyboard.

9.7.2 Selector Switches

So far we have considered binary switches which turn circuits on and off. A manually operated or magnetically operated selector switch connects a single conductor to one of a multiplicity of conductors, as many as 200. Multilevel selector switches do the same simultaneously with several conductors. Figure 9-8 shows electro-magnet 1 which attracts armature 2. Armature 2 is mechanically coupled to pawl and ratchet 3 which rotates one angular step per armature cycle. The moving contacts of switches 4 and 5 are on the same shaft 6 as ratchet 3 so each step of the ratchet moves the moving contacts of switches 4 and 5 from one fixed contact position to the next. Such selector switches are made in many electrical ratings, single directional and bidirectional, single to many switching levels, and so forth.

Some uses of selector switches are for telephone circuits, for circuit changing in electronic instruments, and for connecting power to different windings of a utility transformer. (Telephone switching is now done by solid-state computers of course, but this book includes some obsolete devices both to broaden your understanding and to give you ideas for your own future design use.)

A selector switch may be operated by hand, by an electro-magnet via a pawl and ratchet, or by a motor.

An automobile engine distributor is a selector switch continuously rotated by the engine. It switches sparking voltage from cylinder to cyl-

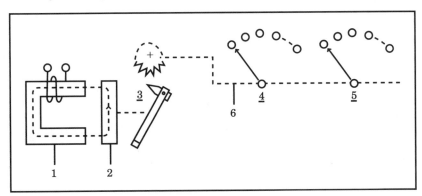

Figure 9-8 Multilevel Selector Switch

inder. Contacts do not actually slide; no metallic contact is made; but high voltage makes the current jump from electrode to electrode over short gaps.

9.8 LIMIT SWITCHES

Limit switches are binary transducers. They indicate when a movable member has passed a particular position on its path; typical movable members are doors and machine slides. Limit switches are packaged with open contacts, enclosed contacts, rugged oil-tight enclosures, vacuum-tight enclosures, and so forth.

A variety of intermediate *operators* are made to couple the sensed motion to the switch contacts; these include long and short levers, sometimes with rollers to prevent sliding abrasion. Limit switches usually have snap-action mechanisms to prevent *teasing* their contacts with gradual and erratic motion which could cause arcing and false signals.

9.9 IGNITION BREAKERS AND BUZZERS

Ignition breakers and buzzers are contacts which open and close rapidly. An ignition breaker on a gasoline engine is mechanically driven by a cam on the engine. It interrupts the current to an induction coil, once per cylinder firing, to generate high-voltage ignition sparks in the cylinders.

A buzzer, Fig. 9-9, has contact 4, driven by electro-magnet 1 and armature 2 working against spring 3, and connected in series with the electro-magnet coil. When the electro-magnet is energized, it pulls the contact open, which deenergizes the electro-magnet, so the device is an EM oscillator. A buzzer is used as a signaling noise maker, sometimes with a small hammer attached which beats on a bell. (The original Ford spark coil used a buzzer with a secondary coil of many turns to generate sparking voltage for a gasoline engine.)

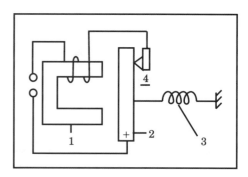

Figure 9-9   Buzzer

9.10 CONNECTORS

9.10.1 General

Connectors are the opposite of relays; they are infrequently engaged and disengaged by hand or by mechanism. They connect one or more circuits (up to 100 conductors in a single connector). Connectors may connect cable to cable, cable to enclosure, or enclosure to enclosure, as when an electronic chassis is plugged into a rack or when two railroad cars are coupled together.

9.10.2 Contacts

Most connector contacts are coital: a rigid male member wedges into an enclosing resilient female member. Most connectors bundle a plurality of such contacts, often of mixed current ratings, into a single pair of holders and engage and disengage the contacts as a group.

The male members may be flat blades, round pins, or printed circuit areas. Flat blades are usually received by flat springs with funnel-like openings to guide the wedging entry. Round pins are usually received by a split tube with a funnel-like entry and with the split segments serving as the contact springs. Printed circuit *edge connectors* are usually mated with flat springs. A connector for coaxial cable uses a central male pin engaging an inner female split tube and an outer hollow male tube engaging an outer female split tube.

Household appliance plugs and sockets and high-current industrial connectors use brass males against bronze spring females. Most connectors for electronics are gold plated to ensure good electrical touching, metal to metal, without intervening oxide films. Wipe is relied on to remove dirt. See Section 9.13 below for a discussion of contact wipe.

In some connectors, the contact pairs butt against each other, each pair having a separate spring. In such a connector, used between subway cars, the individual butting contacts are forced to rotate during engagement to provide wipe.

A large axial force is needed to couple and uncouple connectors having a large number of contacts. Usually an outer ring nut or a pair of screws is provided.

There is a class of *zero insertion force* connectors in which contact force is withheld until after coupling and is then applied by a lever and cam. An experimental connector with 100 contacts was engaged with fingertip insertion force. Then a lever, operated by one finger, applied seven *pounds* of force to *each* of the 100 contacts. The criticism usually leveled at zero insertion force connectors is that they lack wipe during engagement, but the value of wipe, in the presence of high contact force, is uncertain.

A successful undersea connector is shown on Fig. 9-10. The female 1 has a through hole 2 lined with a row of internal contact rings 3; the entire assembly is integrally molded. The male 4 is similar except that it fits through the female with a close fit and has matching external contact rings 5. Both assemblies are greased before submersion. When the parts are mated under water, the grease keeps out the seawater and contact rings 5 touch contact rings 3. The connector is insensitive to hydrostatic pressure.

Among single-conductor connectors are the following:

1. Binding posts for bare wire ends: wraparound the screw types, spring clip types, screw clamp types, and coaxial cable types.

2. Terminal lugs for wire ends, to engage screw terminals. Among these are solder types, solderless crimp types, and screw clamp types for heavy wire.

3. Single-conductor plugs and sockets (or *jacks)* similar to multiconductor connectors but for single conductors.

9.10.3 Bodies

To save space in multiconductor connectors, it is customary to bundle contacts as closely as permitted by voltage stress along the insulator surfaces. Wires are usually soldered into the outboard ends of the contacts because this method consumes a minimum of space. Some industrial connectors use screw clamps instead of solder.

In electronic circuit connectors with slender pins, there is usually mechanical alignment means in the contact holders to prevent people from damaging the pins when they mate the connectors.

Insulating contact holders may be enclosed in a protective metal shell. Some shells are even waterproof. Attachment screws or latches are often provided to prevent pull on a cable from pulling out its connector. A com-

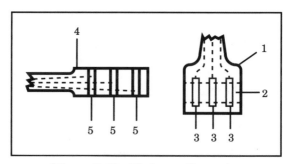

Figure 9-10   Undersea Connector

mon latching means is a ring nut surrounding the connector; another is one or two screws.

A *terminal block* is a set of single-wire connectors packaged into a single assembly. It provides no means for disconnection except for one wire at a time.

9.10.4 Insulation Penetrators

To eliminate the labor of stripping wire ends, insulation-penetrating connectors are used. Figure 9-11 shows a pair of knife-edges, spaced just under the wire diameter, cutting through insulation 2 and indenting and compressing copper 3. It is used with non-stranded wire. This technique is used in telephone installation wiring and in connectors for flat cables. A similar technique uses a sharp cone to penetrate insulation and wedge among the strands of stranded wire.

9.10.5 Wire Wrap

A *wire wrap* terminal is a long, square post with sharp edges. In Fig. 9-12, a stripped, solid wire 1 is tightly wrapped in a helix of several turns around post 2. At each corner, the compression stress is sufficient to penetrate tarnish layers and create a stable, gas-tight, redundant connection. Two or more wires can be connected to each post. Many connections per

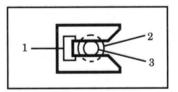

Figure 9-11   Insulation
Penetrator

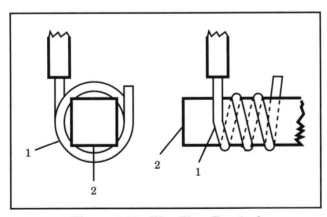

Figure 9-12   Wire Wrap Terminal

square inch of panel area are possible so the technique is much used for computer back planes. Both manual and power wrapping tools are available.

9.10.6 Soldering

Soldering remains a major means to make connections. All connections to printed circuit boards are soldered except for plug-in edge connections and for chip connector sockets which themselves are soldered in. Many connector-to-wire connections are soldered.

Most soldering is done with low-melting-temperature tin-lead solder, but some silver soldering and brazing is done to join metal parts in manufacturing.

Despite their apparent simplicity, connectors account for a large percentage of electronic system failures, so good design and careful use are critical.

9.10.7 Welding

Connections from integrated circuit chips to their lead-in conductors *(lead frames)* are done with short lengths of 0.001-inch diameter gold wire. A wire end is melted into a tiny ball that is then pressed against a gold-plated terminal area of the chip. The pressure deforms the metal which welds as it deforms. The technique is called *thermocompression bonding.* The other end of the wire is similarly bonded to a lead frame conductor.

Fine wire ends are sometimes laser welded.

An experimental electronic packaging system ("Cordwood") spot welded nickel wires together.

Iron rail ends may be thermite welded to produce joints which are good both mechanically and electrically.

Although it is usually considered impossible to spot weld copper, flat wires have been spot welded to printed circuit stripes.

9.10.8 Electrodes

A variety of electrodes are used to make connections to a variety of things. For example:

1. Spring-loaded jaw types: alligator clips, battery clips, spring hook probes
2. Test prods, hand held
3. Graphite electrodes for submergence in liquids (because graphite is chemically inert)

4. Thin sponge pads soaked in electrolyte to make contact with human skin for voltage measurements in electro-cardiographs

5. Noble metal electrodes inserted into the body to measure voltages and to apply voltage from pacemakers

6. pH electrodes of conducting glass immersed in liquids

7. Spot welding electrodes of hard and strong copper alloys

8. Arc welding electrodes of filler metal, or tungsten, or carbon

9. Sparking electrodes of tungsten or molybdenum alloys because of their heat resistance

10. Spot resistance heating electrodes of graphite

9.11 SLIDING CONTACTS

9.11.1 Potentiometers and Autotransformers

A potentiometer uses a sliding brush to contact the turns of its helical winding of resistance wire, that is, to switch from turn to turn. Some potentiometers use a solid resistance element, and the brush touches the selected point along the solid element.

An AC dual of the potentiometer is the EM variable autotransformer. A toroid of copper wire is wound on a ring-shaped magnetic core. The enamel on the wire is removed on the flat face at one end, and a carbon brush is pressed into contact with the bare wire by a spring. The brush is rotated around the axis by hand or motor and acts as a selector switch to choose the desired turn. Carbon is used for the brush so that when two adjacent turns are bridged (as in a commutator), the leakage current is limited.

9.11.2. Slip Rings

Slip rings are sliding contacts used to connect circuits to rotating bodies such as a motor rotor, a cable reel, or a turret (machine or military). A set of slip rings is a set of coaxial insulated metal rings on the rotating body in contact with a set of stationary metal or graphite *brushes* which slide on the rings.

Slip rings may use metal-to-metal contact to minimize resistance and may use several brushes electrically in parallel on each ring to increase capacity and reliability. Slip rings may be configured as short cylinders along a long cylindrical assembly or as coplanar flat rings. Slip rings of different current capacity may be combined into a single assembly.

In instrument synchros, slip rings have given way to transformer coupling between stator and rotor to reduce friction and increase reliability.

9.11.3 Trolleys

Trolleys are sliding or rolling contacts which continue circuits to a body moving along a straight or curved path. Typical trolley applications are electric railways, electric buses, trolley cars (on railroad tracks), and bridge cranes. Long, fixed conductors lie parallel to the path of the moving body, and sliding brushes or rolling wheels are carried by the body. If wheels are used, there must also be slip rings to make connections to the wheels.

The fixed conductors may be:

1. Flexible copper wires hanging from steel cables, as in electric railways.
2. Rigid copper alloy rails as used in bridge cranes. These rails are sometimes called *busbars*.
3. Rigid iron rails on insulated supports as used in subways and some railroads.

Trolleys for crane carriages, machine tool slides, and the like use copper-graphite brushes sliding on insulated copper alloy bus-bars. When travel is less than about 100 feet, there is increasing use of flexible cables instead of sliding trolleys because of the development of hollow-chain cable carriers and festoon cables [14.4].

Railroads use either an overhead copper alloy wire on which slides a transverse brush bar or a soft iron third rail on which slides a spring-loaded cast-iron brush called a *shoe*. In both cases, the return conductor is the railroad tracks on which the steel wheels make contact (only steel-to-steel contact, but at extremely high contact force). A slip ring conducts to the wheels. (Certain lengths of track are insulated and electrically sense the presence of the train wheels as part of the signaling system.) Rubber-tire trains, such as in the Paris Metro subway, have two "third" rails—but no first and second rails.

The brush for a third rail is cast iron. Cast iron has low friction and wear when sliding on steel without lubrication because of the graphite in the cast iron. The brush for a railroad overhead wire, where a contact wheel might come off at high speed, is a transverse bar of very hard graphite, copper impregnated.

Trolley cars and trolley buses use a wheel instead of a brush to contact the overhead wire, again with a slip ring to conduct to the wheel. Trolley cars use their steel tracks as the return path, but rubber-tire trolley buses must use two overhead wires, giving rise to complex and ugly overhead structures, particularly at turns and crossings.

9.11.4 Flexible Cables

Flexible cables are often preferred to trolleys for travel less than about 100 feet. They have greater reliability and less maintenance required by mechanical wear and exposure to corrosion. Flexible cables are supported either by hinged, hollow chain links designed for the purpose or by rolling festoon carriers. One of the advantages of such flexible cabling is that fluid hoses can be included. Section 14.4 describes flexible cable carriers, reels, and festoon systems.

9.11.5 Commutators

Commutators and their sliding brushes are segmented versions of slip rings and are really sliding selector switches, Fig. 9-13. In this case, minimum resistance is *not* desirable; the brush 1, when it bridges adjacent segments 2,3 acts as a load resistor while the current reverses in coil 5 connected between the segments.

Commutators are still made with graphite brushes on copper segments as they have been for a hundred years. The resistivity of the graphite is critical since the brush constitutes a load resistor to absorb the stored magnetic energy of the coil it short-circuits.

9.12 CONTACTS AND ELECTRODES

9.12.1 Switching Contacts

Most switching contacts are made of a noble metal so that corrosion products do not insulate against metal to metal touching. The most common metal is silver, either pure or alloyed. (Silver forms a sulfide in air, but the sulfide is a conductor. Silver is by far the cheapest of the noble metals.) For relays which must operate with low contact force because of little available coil power, gold or even platinum is used. Palladium is an intermediate metal, much used for contacts.

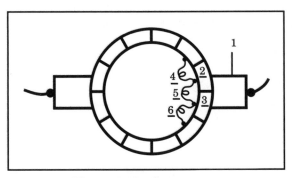

Figure 9-13   Commutator

One of the hazards facing contacts is being welded together by high transient current when they touch. Such transient current, sometimes called *inrush* current, greatly in excess of steady-state current occurs with incandescent lamp loads, capacitor loads, and some motor and transformer loads. The tendency to weld increases if the contacts bounce when they close, which they usually do.

Spot welding electrodes are a special case in which the tips must withstand the very high force used to create the weld; hard copper alloys are used.

9.12.2 Contact Materials

The material parameters important to electrical contacts are as follows:

- Electrical conductivity
- Arc resistance
- Tarnish resistance
- Friction and wear resistance in sliding service
- Hardness and strength
- Contact force (often called *contact pressure)*
- Cost

The noble metals are close to optimum in all respects except cost. Those used for contacts are platinum, rhodium (alloyed with platinum), palladium, gold, and silver. Gold and silver are often alloyed with base metals to increase their hardness.

Silver is the most common contact material for switching devices other than in consumer product switches. Many of these use bronze to save money. Silver is not entirely tarnish resistant; it forms a sulfide film with airborne sulfur compounds, but silver sulfide is a conductor and is easily scraped away when contacts *wipe.* (The tarnish on domestic silverware is silver sulfide.) Silver alloys and mixtures, sometimes made by powder metallurgy, and used for contacts, include:

- Fine silver
- Coin silver
- Silver cadmium-oxide
- Silver tungsten
- Silver graphite
- Silver tin-oxide
- Silver nickel
- Silver molybdenum

- Silver tungsten-carbide
- Silver titanium-carbide
- Silver molybdenum-disulfide
- Silver niobium-diselenide
- Silver nickel tungsten
- Silver copper tungsten
- Silver copper graphite

Some silver alloys are chosen for their resistance to arc damage. When arcing is severe, one or both contacts of a pair may be graphite because graphite does not weld, and its oxide is a gas which dissipates. Silver-impregnated graphite has lower resistivity than pure graphite.

Silver alloy contacts are sometimes diffusion bonded or brazed to a steel, Monel[1], or nickel base layer which in turn is spot welded to the contact carrier. Silver and other precious metal contacts are often made as rivets for installation, the rivet heads constituting the contacts.

The refractory metals, tungsten and molybdenum, have high resistance to arcing because of their high melting temperatures. They are also hard and abrasion resistant. Therefore, they are used where arcing and frequent operation are encountered, despite their high ohmic and contact resistance. High contact force is used to compensate the high contact resistance.

Graphite is unique in that it cannot weld and its oxides are gases which dissipate. Therefore, it is used for arcing contacts, sometimes with one graphite contact mating with one metal contact. Often, a pair of graphite contacts is used in parallel with a pair of metal contacts in such a way that the graphite contacts close first and open last. Arcing takes place at the graphite contacts, and low-resistance, steady-state conduction takes place at the metal contacts.

Graphite has very low friction, so it is used for commutator and slip ring brushes. In commutator brushes, the high ohmic resistance of graphite aids in commutation by providing a load for momentarily short-circuited coils. Graphite can be impregnated with silver or copper for lower ohmic resistance.

A special case is connectors and wires for a thermocouple. To prevent contact potentials from introducing error voltages, each wire and contact must be of the same metal as the side of the thermocouple to which it is connected. Each junction is itself a thermocouple in series with the working thermocouple.

Another special case is a long comb of contact springs used to block RF leakage through the crack of a closed door of a shielded enclosure.

[1] Registered U.S. Patent office.

Usually, these contacts are merely the beryllium copper of the spring strip itself.

Excellent electrical contact is made between parts made of iron, steel, and stainless steel which are pressed together with very high forces for mechanical purposes, as in railroad wheels on their tracks.

9.12.3 Electro-Plated Contacts

Connector contact areas are often electro-plated with gold. This provides a reliable connection at low cost, provided that the number of engagements and disengagements during the connector's life does not wear off the gold and provided that the contact is not subject to arcing by being used as a switch.

9.12.4 Sealed Contacts

The atmosphere is an enemy of contacts. It carries corrosive chemicals and dirt, its oxygen burns contact material, and it ionizes and conducts arcs.

To protect some contacts from the atmosphere, they are hermetically sealed in either vacuum or inert gas. If the entire device were so sealed, the cost would be higher, heat dissipation would be extremely difficult in vacuum, and coil outgassing might foul the enclosed space. Therefore, the usual practice is to seal the contacts but to couple them mechanically or magnetically to the outside electro-magnet which operates them. However, some miniature relays are entirely hermetically sealed in inert gas.

A first way to couple an actuator to a sealed contact is to provide a flexure—a bellows or a flexible diaphragm—as part of the sealing enclosure. The actuating electro-magnet exerts force on the outside of the flexure, and the inside of the flexure exerts force on the contact. If the enclosure is pumped out to a high vacuum and if the contacts are of refractory metal, arcs either do not form at all or are not sustained; such contacts are highly resistant to arcing.

9.12.5 Mercury Contacts

Mercury is unique in being a metal which is liquid at room temperature. When it is enclosed in a sealed container and displaced by gravity or otherwise, it can make contact with electrodes which penetrate the container. Enclosed mercury can also be used as both contacts of a pair, and their contacting is usually bounce free.

The mercury may be displaced by tilting the container so gravity moves the mercury. It may also be displaced by penetrating the container with the magnetic field of a coil which pulls down an iron armature floating on the mercury, thereby raising the level of the liquid.

9.12.6 Reed Switch

An enclosed pair of magnetic cantilever springs, the *reeds,* lie in the magnetic circuit of a coil. When the coil is energized, electro-magnetic force draws the reeds together. The touching areas are electro-plated with noble metal and provide a reliable contact for low-voltage and low-current circuits having no arcing. Reed switches can also be activated by moving permanent magnets on mechanisms, so they are also used as limit switches and selector switches.

Similar to the reed switch is the mercury wetted switch. It differs from the reed switch in that it contains a small pool of mercury. The mercury is carried to the contact area by capillarity along a split in each contact arm. The contact is mercury to mercury. The contact is bounce free and long lived so the mercury wetted switch is used in high-speed relays transmitting information codes and for dry-circuit switching.

9.13 CONTACT BOUNCE, WIPE, AND GEOMETRY

It is conventional wisdom that contact force should be concentrated on a small area so that true contact pressure—force per unit area— be high. High contact pressure breaks through the dirt and tarnish on the contact surfaces to establish true metal-to-metal touching. (It is an unfortunate convention to use the phrase *contact pressure* for what is really *contact force.*)

Force concentration is achieved by rounding one or both contacts of a pair into a domed shape or by making both with edges that cross each other to achieve a nominal point contact.

There is a further benefit from such localization of contact touching. Contact dimensions are often substantial compared to the dimensions of their supporting structure. If the point of touching of nominally flat contacts were near an edge instead of at the center established by the rounded shape, the distribution of forces among several contacts might be substantially different from the distribution desired.

There is a contradictory argument, however. All surfaces have some roughness—peaks and valleys—and two pieces of metal can actually touch only peak to peak. (The peaks are called *asperities.*) Therefore, regardless of apparent contact geometry, there is a microgeometry of the contact surfaces which confines actual metal-to-metal touching to one or more quite small areas, regardless of the designer's intent. Current concentrates at these small areas of actual connection. It is because of these current concentrations that high ohmic conductivity is desirable in contacts.

Another technique to achieve a reliable connection is to connect two contact pairs in parallel, each individually spring loaded. The probability of both pairs failing to connect is much less than the probability of a single pair with twice the contact force.

When butting contacts close, they bounce like billiard balls colliding. Electronic circuits may have to filter out the multiple closings and openings to avoid errors. Power circuits may arc between bouncing contacts, and such arcing not only burns away contact metal but may cause the contacts to weld together when they reclose on the arc.

Contacts engage each other by butting, by sliding, by rolling, and by wedging. Most relay and contactor contact pairs butt against each other like rams in conflict. Trolley contacts slide or roll, but rolling trolleys must have a sliding contact in series to make connection to the roller. Commutator and slip ring contacts *(brushes)* and some selector switch contacts slide. Most connector contact pairs wedge, as do the contacts in large circuit breakers. *Wiping* contacts on relays first butt and then slide.

Where the cost is justified, bounce-free contacts can be made. In one system, the momentum of the moving contact is transferred through the fixed contact to metal powder whose momentum is absorbed by friction and many fine impacts. In another system, the impact momentum is transferred through the impacted stationary contact to a third body, of equal mass, which then dissipates the momentum, with or without bounce, against a final stop. (Remember what happens when a billiard ball hits two touching balls in the line of motion.) Another design approach is to use wedging contacts. Mercury contacts do not bounce. Electric organs use flexible springs as bounce-free contacts.

Carbon and high-melting-point metals such as tungsten and molybdenum are used to prevent welding due to arcs induced by bounce. Sometimes, two contacts in electrical parallel are used, the first close to being weld resistant and the second being of higher conductivity. Contact welding is the obverse of the circuit breaker problem of resisting arc damage when the contacts separate.

Many—but not all—relays and contactors have kinematics such that their contacts rub against each other for a short distance *(wipe)* when they come together in order to scrape off any insulating layer which may have formed. And many do not.

Sensitive relays do not have enough mechanical power to wipe or to generate much contact pressure so they rely on gold or even platinum contacts. Non-wiping contact makers rely on high contact pressure and noble metals to obtain reliable contact.

Optimum contact design is not yet an exact science.

9.14 DRY CIRCUITS

Circuits having low voltage and low current are uniquely difficult to close. Only gold, platinum, or palladium contacts can be used.

9.15 ARCING AND QUENCHING

In most devices, the contacts operate in air. This exposes them to tarnishing, to dirt deposit, and to burning from arcs. Arcing is the primary enemy of contacts and contact-making devices. In circuit breakers for high-voltage and high-current circuits, it is extremely difficult to extinguish the arc at all and there is an extensive technology for doing so, which we have seen described in Section 9.4.

9.15.1 Welding, Arcing, and Sparking

So far, we have considered devices in which the contacts should have as low resistance as possible. A *spot welder* joins two pieces of metal by the heat generated by resistive power consumption in the contact between the two pieces. Low resistance prevents such power consumption; copper, silver and gold cannot be spot welded at all. Furthermore, contact resistance between a spot-welding electrode and its workpiece does no harm; in fact, the additional heat helps out. However, the electrode must withstand high contact force and must not, itself, weld to its workpiece. A copper alloy is usually used for the electrodes.

Seam welding is spot welding in which the electrodes are rolling wheels which produce a closely spaced row of spot welds.

(Dielectric welding of plastics resembles seam welding except that the workpieces are insulators and the electrodes do not conduct current to them but introduce it as a high-frequency electric field. Any metal will do for the electrodes.)

Electric arcs are used for welding as well as for lighting. A momentary electrical contact must be made to initiate the arc. If a filler metal rod is not the electrode, the electrode must withstand the extreme heat of the arc; usually, this electrode is of tungsten or of graphite. It is common practice to cover the arc with flowing inert gas to minimize burning. In some automatic welders, the arc is submerged in a powder which gives off a protective gas.

Lasers are used to apply heat for evaporative cutting, for burning, and for welding. The heat can be used with any material. Laser heat can be focused onto extremely small areas.

Sparks are used to ignite the fuel in gasoline engines. Although physical contact is not made between the spark plug electrodes, they must withstand erosion by the spark and the forces, temperatures, and chemicals of the burning gasoline.

In *electrical discharge machining* (EDM), an electric current in a liquid between a tool and a workpiece erodes the workpiece into the form of the tool. The tool may be the negative of the finished part or it may be a wire analogous to a band saw blade. The action is partly electro-plating and partly arc erosion.

Ultrasonic machining resembles EDM except that the tool is vibrated at ultrasonic frequency by a piezoelectric crystal and erodes the workpiece with a slurry of abrasive particles. The cutting action is not electrical.

9.16 FUSES

A fuse serves the same function as a circuit breaker, but it interrupts a circuit by melting instead of by opening a contact. It must be manually replaced after each operation.

A fuse is the electrical analog of a mechanical shear pin; when an overload occurs, the fuse link melts and interrupts the circuit. Most fuses are of low-melting-temperature alloy, but some high-voltage fuses are of silver embedded in sand. The cross section of the arc is small, and the arc is in intimate contact with the cooling sand. Fuses are usually made integral with a two-contact connector so that they can be replaced easily.

Fuses have no moving parts, require no maintenance, and have unlimited life before blowing, so they are extremely reliable. On the other hand, reset after blowing is time consuming, and a replacement fuse of the correct rating may not be readily available.

Some *slow-blow* fuses have a thermal capacitance added so they do not melt on transients, a feature that shear pins lack. However, since most electrical equipment failures are due to overheating, which takes time, and most mechanical failures are instantaneous, the lack of time delay in shear pins is usually a good thing.

Many fuses have a window so that the melted link can be seen as an indication that the fuse has blown.

9.17 LIFE

The required life of electro-mechanical contact-making devices varies from only a few hundred cycles for most connectors and circuit breakers to hundreds of millions for data-transmitting relays. DC-operated devices tend to outlive AC-operated devices because they have less shock when operated.

9.18 OTHER DEVICES

There are many other devices which use electrical contacts.

One is mechanical go-no-go gauges in which closing of a circuit between an electrode and a workpiece, or between two electrodes, indicates whether or not they actually touch.

Another is a combination adjustable hard stop and limit switch. The same stainless steel parts serve both as stops and electrical contacts. Stainless steel is one of the worst contact materials one could find because of its surface oxide film, but with 75 *pounds* contact force from the actuating air cylinder, it works just fine!

Metal-to-metal contact making is a technique which invites its use.

CHAPTER 10

Computer Components

It might be said, in current slang and with some exaggeration, that electronic computing is not all that electronic. This chapter surveys the electro-mechanical devices that make the actual computing, which *is* electronic, practical and useful.

10.1 MANUAL INPUTS

10.1.1 Keyboards

A part of a computer is an EM keyboard which transduces mechanical motions by a human into closings of electrical circuits. The keys are limit switches with labels. (Some computers receive inputs from non-human sources.) In addition to traditional typewriter keys, there are function keys and duplicate numeric keys.

A kind of keyboard uses diaphragm contacts (see Fig. 9-7) labeled with special functions such as machine tool control codes or the components of a restaurant order.

Another kind of keyboard is not hardwired to the computer, although the keys themselves are the same kinds of limit switch as above. Either it is coupled in real time via radio or it has a solid-state memory packaged with it; the memory is plugged into the computer from time to time for data transfer. For example:

1. Waiters use the radio-coupled keyboard to speed up service.
2. Military observers use the radio-coupled keyboard to transmit coded reports instead of using voice radio.

3. Electricity and gas meter readers, package delivery people, and inventory takers use stored-data keyboards.

Another kind of keyboard resembles a stenotype machine or a small set of piano keys. It is used on some mail-sorting machines to permit a human address reader to tell the sorting machine each letter's destined sorting bin by a single stroke or a combination of keys.

Touch screens, Section 10.1.4 below, are a kind of keyboard.

Telephones, calculators, and control panels for burglar alarms, thermostats, and the like have keyboards, mostly of the limit switch type but sometimes of the diaphragm type.

Electric organs use limit switch keyboards with bounce-free contacts.

10.1.2 Voting and Scoring

A voting system provides a standard tab card and a stylus. Each hole is partly punched through, and the voter uses the stylus to complete the punching. A labeled template identifies the holes. Standard card-sorting machines count the holes.

In a test scoring system, a person marks a multiple choice location with a soft graphite pencil. A machine senses the electrical conductivity of the marks.

10.1.3 Light Pens

A light pen is a wand with a photoelectric cell at its tip. A human touches the computer screen with the tip at a location marked in the display. At some instant in each screen-writing cycle, the photoelectric cell is illuminated. The computer determines the location from the time of illumination.

A wand having both a light source and a photoelectric cell reads a bar code when the tip of the wand is wiped along the bar code.

There are automatic bar code readers, the most common being at supermarket checkout stations.

10.1.4 Touch Screens

A transparent screen covers the computer cathode ray screen. When the transparent screen is touched by a finger the coordinates of the touch point are measured. The measurements are transmitted to the computer, which relates them to the defining display. In effect, the user touches a displayed push button.

A fine-resolution touch screen reads hand-lettered alphanumeric characters.

Several techniques are used to make a transparent touch-sensitive screen. Among them are:

1. A resistive film is alternately energized from top to bottom and left to right. A flexible conductive film overlies the resistive film with a small air gap. The user's finger presses the conductive film against the resistive film, the point of contact becoming, in effect, the wiper of a two-axis potentiometer.

2. A resistive film is energized by high-frequency AC applied to a pattern of electrodes around the edge. A finger provides a capacitive coupling to ground at its touch point. Voltage measurements of the electrode array indicate the touch point.

3. Surface acoustic waves are interrupted by the touch and the timing of the interruption indicates the position of the touch.

4. The entire monitor is mounted on strain gauges. The touch force adds to the strain gauge loads and its position is computed from them.

And there are others.

10.1.5 Plotting Digitizers

The user places crosshairs on individual points on a drawing on a digitizing table. When he presses a button the magnetic coupling between a coil around the crosshairs and wires embedded in the table, with an analyzing circuit, digitizes the X and Y coordinates of the crosshair position and informs the computer.

10.1.6 Image Scanner

The user slides a linear array of photoelectric cells across a black-and-white image such as a signature. A raster of points is memorized and played back on call.

10.1.7 Magnetic Stripe Credit Cards

The user slides a credit card along a guide groove. A magnetic head reads the codes on the card. In prepaid train fare and telephone cards, a second head rewrites the magnetic stripe with the credit remaining after the current transaction.

10.1.8 The Mouse

The *mouse* is the most widely used manual input device after the keyboard. It is a hand-size box in the bottom of which is a loose ball that rests on the table. When the box is moved along one axis, the ball rolls in that direction and generates a stream of pulses to the computer on a first channel, and when the box is moved along the perpendicular axis, the ball rolls in that direction and generates a stream of pulses to the computer on a second channel. Rolling in an intermediate direction generates both streams of pulses in response to the components of the motion.

Software converts the two streams of pulses into corresponding motion of an arrow on the screen, or, in computer aided drafting programs, a pair of coordinate lines moves on the screen. One, two, or three auxiliary push buttons on the mouse command computer responses related to the position of the arrow or of the coordinate lines.

Mouse action is used to select icons, or to select command words, or to "drag" portions of the display, or to perform drafting operations.

A *track ball* is an inverted mouse; the user's hand rolls the ball directly.

10.2 MAGNETIC RECORDS

10.2.1 The Magnetic Circuit

In Fig. 10-1, magnetic core 1 carries coil 2. Current in coil 2 from a *write amplifier* induces a magnetic field 3a,3b. Portion 3a of the field crosses air gap 4 and has no useful effect, but a fringing portion 3b passes

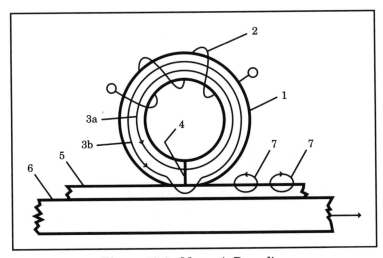

Figure 10-1   Magnetic Recording

through magnetic medium layer 5 and magnetizes a portion of it, leaving a record 7. Magnetizing in one direction corresponds to a 1 bit, and magnetizing in the reverse direction corresponds to a 0 bit. Magnetic medium 5 is carried by mechanical substrate 6 kept in motion relative to core 1 so that a stream of bits is recorded.

When the motion is repeated with the coil connected to a *read amplifier* input instead of to a write amplifier output, each magnetic bit record sends magnetic flux back through the core and coil and induces a voltage fed to the read amplifier. The core and coil are called a *magnetic head.* Magnetic tapes may have separate read and write heads with slightly different characteristics.

Substrate 6 may be a flexible Mylar ribbon or *tape,* a flexible Mylar *floppy disk,* a rigid aluminum *hard disk,* or a plastic or paper card carrying a short stripe of magnetic medium 5. (The earliest flexible medium, now obsolete, was magnetic wire.)

With tape, floppy disk, and striped card, the head is in physical contact with the magnetic medium so there is no magnetic reluctance in an intervening air gap. The spacing between bits is determined by the width of gap 4. With a hard disk, the mechanical speed is much higher, and the magnetic head rides, or floats, on a hydrodynamic bearing comprising a film of air between the head and the medium. The spacing between bits is determined by both the width of gap 4 and the thickness of the air film.

In the early development of magnetic recording, before floating heads were developed, hard drums were used instead of hard disks. Heads did not float but were mounted on a rigid structure with a very close clearance to the medium surface.

Cores 1 are made of low-hysteresis ferrite. Recording medium 5 is made of high-hysteresis ferrite powder mixed with adhesive that bonds it to the substrate. Air gaps 4 are made extremely thin in order to shorten the length of the bit record and thus increase the number of bits per inch.

Figure 10-1 shows a single head and a single track. Usually, the bit format on computer magnetic tape is one byte of eight bits side by side; correspondingly, eight heads are stacked side by side. The bit format on both floppy disks and hard disks is serial, and only a single core is used, but there are many concentric record tracks, and the head is quickly shifted radially from track to track by a voice coil actuator. For hard disk recording, cores 1 are made as lightweight as possible to permit rapid shifting and to minimize the load on the hydrodynamic bearing.

10.2.2 Tape Transports

The mechanism, or *transport,* for the magnetic tape used to back up a computer's memory resembles an audiotape transport except that the audiotape is moved at constant speed and the computer tape may be driven back and forth to go to the desired segments.

The transport for a tape used as a random-access memory (RAM), quickly moves the tape to random locations chosen by the computer. The transport's speed limits the computer's speed. The mechanism is much more complex than that for a continuous-motion audio or backup tape, as shown in Fig. 10-2:

Tape 1 is transported back and forth between reels 2,2a by low-inertia reversible capstan 3. It is read and written on by head assembly 4. The inertia of the loaded reels is far too great for their motors to accelerate the tape at the desired rate, so buffer storage is provided for the tape. Vacuum wells 5,5a are pumped down by vacuum pump 8. Thus the capstan need only accelerate half the length of tape in the wells and the reel motors need only accelerate enough to keep the tape loops within the tolerance of the well depth. The position of the tape loops in the vacuum wells is sensed by vacuum switches 6,6a at the high end and 7,7a at the low end of the wells. When the tape covers a switch, the switch pressure rises to atmospheric and the switch turns its motor on or off.

An alternative to vacuum wells are *dancer rolls* such as used in metal and paper manufacturing. The buffer loop is tensioned by one or more dancer rolls on a spring-loaded swinging arm.

10.3 MAGAZINES

All tapes and replaceable disks require EM means to store, insert, and replace the tape or disk. Tapes are stored on reels that may be in cassettes, floppy disks in either paper or plastic envelopes with head-access windows, and hard disk *diskpacks* in mechanical stacks. The most elaborate tape-handling system is used in videotape recorders (VCRs) in which the user merely pushes a cassette into a hole, a fully automatic EM threading operation takes place inside the hole, and the hole's mechanism later hands it back.

10.4 CD-ROMs

Audio compact disk (CD) recorders sample the sound to be recorded, digitize each sample, and record the binary digit stream in a spiral of reflecting or non-reflecting spots on the disk. The digits are read by reflecting or not reflecting a very small laser spot that follows the spiral. The bit density is very much greater than the bit density on magnetic recording disks, so much more information can be stored on the same area. However the memory on such a disk cannot be changed, so it constitutes a *read-only memory,* or ROM. Needless to say, R&D on RAM CDs is under way. CDs were originally developed for audio recording but were soon adapted to computer data recording.

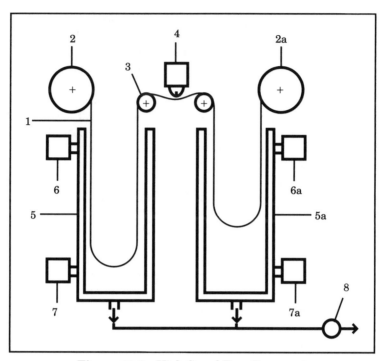

Figure 10-2   High-Speed Tape Transport

10.5 OPTICAL CHARACTER RECOGNITION (OCR)

Printed characters are read photoelectrically and the information is fed into the computer in binary code. Section 7.4.10 describes the operation.

10.6 PRINTERS AND PLOTTERS

Chapter 11 covers the great variety of printers and plotters that deliver computer output.

10.7 ACOUSTIC DEVICES

10.7.1 Sound Output

Computers have long had loudspeakers to deliver warning signals. They now have the audio electronics to deliver speech, music, and sound effects. CD-ROMs can record such sounds for the computer to deliver, and the computer can synthesize sounds, including speech.

10.7.2 Sound Input

Computers can recognize speech to an increasing degree and respond to it.

10.7.3 Delay-Line Memory

In the early development of digital computers, before large and fast magnetic memories were developed, a number of other memory techniques were used. One of them used a long tube filled with mercury. Binary code pulses of sound were introduced at one end by a loudspeaker, detected at the other end by a microphone, amplified, and returned to the source end. The memory was the stream of sound pulses moving along the tube between the loudspeaker and the microphone. The device was called an *acoustic delay-line memory;* it is now obsolete.

10.8 SIMULATORS AND TRAINERS

Imagine an airplane cockpit minus the rest of the airplane and with a student pilot in the cockpit. The cockpit is mounted on a tilting mechanism that turns it in roll, pitch, and yaw in response to a computer's commands; the simulator's instruments exhibit data commanded by the computer; and the cockpit control switches and levers generate inputs to the computer.

Now assume that the computer is programmed to calculate the behavior of a real airplane responding to the student pilot's actions and also responding to signals from a human teacher who tells the computer that there is a storm, that the airplane is landing, that an engine has caught fire, and so forth.

The computer calculates the performance of the imaginary airplane in response to all its inputs. It generates signals to the cockpit instruments and to the cockpit's EM support mechanism so that the student pilot is in the same situation as he or she would be in a real airplane (even sound effects). (An exception is linear acceleration, which would take a lot of space!)

This system is a *training simulator;* it is faster, cheaper, and safer than practicing with the real hardware it simulates. Training simulators have been developed for many kinds of machines and situations, from relatively simple machines for training maintenance technicians to complicated battlefields for training generals.

Simulators are also used to perform experiments in the development of new machines. They may simulate the entire machine or may be connected to portions of the machine made of real hardware and simulate the remainder of the machine.

Simulators originally used analog computers, then hybrid analog and digital computers, and now digital computers. The evolution was caused by the slow evolution of analog computer technology and the incredibly fast evolution of digital computer technology.

The latest wrinkle in simulators has been named *virtual reality* by its promoters. It adds stereo images, usually in TV tubes in front of the student's eyes or else in holograms, both with computer-generated images.

(A manned spaceship simulator was built in the early 1960s. It used a variety of EM models and real TV cameras looking at them to generate the moving images. Now this is done by software.)

10.9 DISPLAYS

Most computer display is by cathode-ray tube or light-emitting diodes (LEDs) but there are EM displays which are advantageous in some circumstances. Most are arrays of painted dots, characters, or words that are selected and displayed by EMDs under computer control. Most use ambient light for illumination. Among these are:

1. Matrices of large dots

2. Tapes, drums, or strips bearing characters or words and positioned by EM actuators

3. A matrix of microscopic mirrors etched from a silicon chip, each mirror with a voltage-sensitive silicon actuator. The "1" mirrors reflect light to a display panel and the "0" mirrors do not.

Matrix displays are also made with LEDs or incandescent lamps instead of mechanical dots. These displays are easily made as traveling signs. An old one is preelectronics: the traveling sign on the New York Times Building in New York City. Originally, it was controlled by arrays of sliding EM contacts. Uses include roadside signs, stock price displays, and railroad schedule displays.

Not EM at all, the simplest changeable displays still in use, using electricity only for backlighting, are the hand-cranked destination signs in buses and the hand-placed letters on movie marquees.

10.10 HEAT REMOVAL

Every electrical device generates heat that must be removed or the device will destroy itself. Where available space and ventilation are plentiful so a circuit can be spread out, heat is removed adequately by natural convection, conduction, and even radiation. However, in computers, conductor lengths must be minimized to minimize signal propagation time. Therefore, special means must be provided to remove heat.

Among such means are motor-driven fans to circulate outside air over heat sources, refrigerated air or other gases or vapors in the electronics space, and massive aluminum heat conductors. This subject is treated at greater length in Chapter 14.

Certain computer circuitry is much faster at cryogenic temperatures. If this technology is developed to a commercial level, EM refrigerators for very low temperatures will be required.

CHAPTER 11

Marking Paper

We all know that computers have made possible the "paperless society" and have also made possible the enormously increased output of paper which has actually occurred. This chapter surveys both computer-driven and non-computer-driven kinds of marked paper and the EMDs for marking and paper handling. It includes:

1. The variety of end products
2. The kinds of paper
3. The kinds of marking process and the devices which use them

11.1 VARIETY OF END PRODUCTS

The following list includes most kinds of marked paper, but it does not include many artistic and industrial variations. Please think of the list extended to include combinations of these items.

1. Black and white text, with different degrees of sharpness, contrast, and clarity
2. Single- or multiple-font text
3. Color: black on white, two color on white, multicolor on white, colors on colored paper
4. Graphics, solid color or shaded, as in drafting
5. Images in shades of gray and in shades of color
6. Drawings with all of the above options, plus large size, plus translucency (to make contact prints)

7. Photo reductions, from contact prints down to microimages and up to enlargements

8. Quantities from single copies through multiple copies to mass copies

9. Marks on curved surfaces, including non-paper

10. Marks on preprinted graph paper or on business forms

11. Marks on specialty papers, described in the next section.

11.2 KINDS OF MEDIA AND THEIR MARKING MEANS

This list is confined to the different kinds of paper intended to receive images from different sources; it does not include the myriad of papers designed primarily for aesthetic effect or mechanical functions such as packaging. Some of these papers were developed jointly with the marking processes described in the next section. (This would be a good time to reread Section 2.11.2 on lithography.)

1. *Plain paper* has no coating to make it sensitive to a particular marking process other than applying ink. It does not have the glare, curl, flimsiness, or other undesirable attributes of some papers designed to permit certain of the processes described below. Many copiers are advertised as using plain paper. "Plain paper" includes drafting vellum and plastic film capable of holding ink. There is a great variety of weight, surface finish, color, component fibers, and surface texture.

2. *Electro-sensitive paper.* Electric voltage from an electrode leaves a mark.

3. *Thermal paper.* Heat leaves a mark.

4. *Photochemical papers and films.* These have a coating of chemicals. They receive a light image which makes a first chemical change, and most are then processed chemically to make a permanent visible image. They vary from blueprint or diazo papers through high-resolution photographic films, both black and white and color.

5. *Impact sensitive paper.* These are used to make multiple copies from an impact printer. Typically, the underside is coated with material which releases ink when the face is struck and the ink marks the face of the adjacent layer. Carbon paper (which no longer uses carbon) is a form of impact-sensitive paper which is then discarded. Impact-sensitive paper is usually also sensitive to non-impact pressure from ballpoint pens used to add signatures and from printing type.

6. All the above papers which are preprinted with graph lines or business forms or adhesive areas, or which are die stamped with tearing perforations or sprocket holes, or which are preformed into envelopes or other shapes.

7. All the above papers which are either cut sheets, rolls, sprocketed fanfolds, or ribbons.

8. *Tabulating cards.* Tab cards are heavyweight plain paper. They are marked both with ink and with punched holes.

9. *Teleprinter tape* of paper, fiber, or Mylar. The tape is marked with punched holes and sometimes also with inked characters. The holes are intended for mechanical reading with either spring-loaded pins, air jets, or light beams.

10. Letters in envelopes (ready to have their stamps canceled and then sorted).

11. Envelopes and labels to be addressed.

12. Magnetic tape, disks, and cards, while not human readable or really paper, are recording media which are marked magnetically and are mentioned here for comparison. Some cards are really paper.

13. Readable displays which do not actually use paper but provide a changeable display of colored chips, printed strips, belts, and drums, and so forth. Most of these non-TV or lamp displays are EMDs; see Section 10.9.

14. Very early laboratory recorders used a smoke-covered drum turned by clockwork and a sharp stylus which scratched a record.

11.3 PAPER HANDLERS

At this time, all paper marking by machine is indelible, and the paper is not reused, so a mechanism must be provided to feed fresh paper from a magazine. In many color processes, the same piece of paper is written on again for each color so it must be rehandled.

In some processes, the paper sheet is stationary during two-dimensional marking; in some processes using a rolling printing drum, the paper moves synchronously with the drum; and in some plotting processes, the writing element moves in one direction and the paper moves in the perpendicular direction. (In flat-bed letterpress printing, the printing plate merely moves toward and away from the paper.)

Cut sheets are provided in a stack from which single sheets are removed without disturbing the stack and without removing more than one sheet at a time. Usually, feed rollers are used which have more friction with the top sheet than the top sheet has with the second sheet. Sometimes the paper is flexed to separate the sheets. Sometimes vacuum is used to provide more force between the driving friction member and the top sheet than between the top sheet and the second sheet. Sometimes vacuum cups lift off the top sheet.

Tab cards are fed from the bottom of the stack by a pusher which is thinner than a card. (As was said, this book mentions obsolescent technology too.)

The ultimate in simplicity and reliability is feeders for sprocketed fanfold paper, but the secondary operation of separating the sheets and also their rough edges are disadvantages.

A very elegant and successful pseudosprocketing is done in some pen plotters by pinching the paper edges between feed rollers coated with rough grit. Each grit particle embosses a tiny pit in the paper. The grit particle registers mechanically with its pit each back and forth pass, rather like a tiny sprocket pin.

Multiple-pass processes for color printing or pen plotting may depend for image registration on mechanical guidance of the paper, such as by sprockets, or may use photoelectric sensing and servo control, as is done with color printing of newspapers, books, magazines, wallpaper, and cloth. Alternatively, the paper may be gripped to a stationary platen by vacuum and the pen driven on both axes.

Paper of all types is conveyed through marking machines by sets of rollers.

11.4 IMPACT SERIAL PRINTERS

Paper markers may be divided between serial and parallel. Serial printers mark paper one character at a time or, with a stylus, one point at a time; typewriters and strip-chart recorders are serial devices. Parallel markers mark paper along an entire line or over an entire area at the same time; for example, photographic devices are parallel devices. Dot matrix printers mark a single column of points in parallel and a series of columns in series.

11.4.1 Character Printers

Typewriter usually refers to a hand-operated machine which prints one character each time a person strikes a key. *Teleprinter* extends the category to include EMDs with signals coming from a telegraph line, computer, cash register, or the like. A font of raised characters is poised over the paper, and one character at a time is selected and strikes the paper through an inked ribbon. The paper is supported by a cylindrical rubber *platen* which serves both as an anvil for the impact and as a feed roll. In some, duplicate fonts are arrayed in parallel rings on a drum, and a traveling hammer strikes one character per ring as the drum is rotated to the desired position for each character.

The font may be:

1. On a set of independent levers as in the original typewriters

2. On a cylindrical drum or spherical barrel which is positioned in two axes to make a selection

3. On a single ring of flower-like petals, one character per petal, which is rotated about a single axis (a *daisy wheel)*

4. A *dot matrix* of individually operated pins which synthesize any desired font [11.4.2]

5. One computer terminal used a font in single file on a chain which was driven along a printing line; a separate hammer per character position struck the chain when the desired character was over its position.

6. An early hybrid printer/plotter made by Typagraph Corp. used stepping motors to move the paper up and down on a typewriter platen and the font barrel left and right. Resolution was 40 positions per inch, the spacing which made adjacent periods just tangent to each other. The machine not only printed ordinary text, but it typed graphs as sequences of closely spaced characters, typically periods. It then typed coordinate axes and labeled the graphs with typewritten characters.

11.4.2 Dot Matrix Printers

The dot matrix printer is the most common computer printer at the time this is being written. It uses a raster or matrix of closely spaced dots to form each character. A dot is formed by a dot module, a single wire extended by an electro-magnet and impacting the paper through an ink ribbon. A column of dot modules moves sideways across the paper. Each dot module electro-magnet is energized momentarily when its dot forms part of a desired character. This takes a lot of control logic, but electronic logic is cheap. Furthermore, no particular font is built into the machine. It can produce as many fonts as the computer has programs; in fact, it can produce large characters by building them up with two or more passes. Similarly, a dot matrix printer can be used as a graphics plotter.

11.5 NON-IMPACT SERIAL PRINTERS

11.5.1 Laser Printer

A laser printer uses the same photosensitive drum and toner process (xerography) as the electro-static photocopy machine [11.7.3] except that the optical image imposed on the drum is produced by a sharply focused laser beam which scans the drum in closely spaced lines while the laser is rapidly turned on and off. Geometrically, it is equivalent to the dot matrix printer above except that the resolution is very much finer and the speed is very much faster. If the same piece of paper is processed several times, each time with a different color toner, a multicolor print results.

11.5.2 Ink-Jet Printer

An ink-jet printer is analogous to the mechanical dot matrix printer above except that the wire is replaced by a stream of fine ink droplets turned on and off at high speed; each droplet is propelled by a pulse of electrically generated heat. It can be used on any surface to which ink will adhere; there was once an ad which showed it writing on the yolk of a raw egg without breaking it.

11.5.3 Plotting Printer

Any of the serial X-Y plotters described below can be programmed to draw alphanumeric characters of any size or font. This is too slow a process for text printing but it is satisfactory for labels on plotted drawings.

11.6 PLOTTERS

11.6.1 Strip-Chart Recorder

A *strip-chart recorder* plots one or more parameters as a function of time. The *strip* is a ribbon of paper moved through the instrument at uniform speed, usually by an electric motor [Fig. 6-2].

A stylus moves back and forth across the strip in proportion to the variation of the recorded parameter. The stylus is usually at the tip of the needle of a large d'Arsonval galvanometer. The stylus may be a hot wire marking heat-sensitive paper, or an ink pen, or an electrode marking voltage-sensitive paper.

Strip-chart recorders are also made with several styli simultaneously recording several parameters on a wide strip of paper.

For higher-frequency signals, a tiny galvanometer coil carries a tiny mirror; a reflected light beam leaves a record on a moving strip of photographic film [Fig. 6-3].

A still higher frequency system comprised a cathode-ray tube with a bundle of optical fibers embedded in its face. A photographic film was dragged across the external end of the fibers. Without the fibers, the light would leave only a blurred trace because of the thickness of the glass between the phosphor light-emitting surface and the film surface.

For very high frequency signals, a camera photographs the screen of a cathode-ray tube.

It is common to build strip-chart recorders with two or more channels side by side so that the relative variation of two or more parameters can be recorded. Mirror galvanometers are typically arrayed side by side to make simultaneous records on the same photographic film.

Circular recorders are the polar-coordinate equivalent of the Cartesian coordinate recorders we have considered so far. The chart is a disk, and the parameter is recorded as a radius on the disk. A typical use is to drive the disk at one revolution per day, changing it daily, to record a slowly changing parameter such as temperature.

Non-EM chart recorders are made with clockwork or other mechanical paper feeds and with piston and cylinder, bellows, bimetal, float and string, or other non-electric stylus drives. For example,

The *steam engine indicator* uses a paper-covered drum rotated back and forth (X-axis) by the reciprocating motion of the engine piston and a pencil stylus driven parallel to the drum axis (Y-axis) by cylinder steam pressure. The resultant plot shows pressure vs. piston position. Totally non-EM!

11.6.2 X-Y Plotter

X-Y plotters are used to generate engineering drawings, maps, and the like. They are made for paper sizes from notebook to many feet by many feet.

Plotters were originally recorders of real-time analog signals just as strip-chart recorders still are. The difference is that the abscissa could be something other than time. Then came computer-generated mathematical functions and then came Computer Aided Drafting (CAD). Most large X-Y plotters are now the hard copy drawing boards of CAD systems.

Plotter paper may be fixed to a flat bed, fixed to a rotary drum, or fed as a belt of unlimited length by small diameter rollers.

Flat-bed plotters have stationary paper, and the stylus is driven in both X and Y coordinates. One interesting use of the larger sizes is to draw clothing patterns; in fact, a computer is used to nest the different shapes as closely as possible to minimize wasted cloth.

The distinction between drawing and fabricating has been blurred by the use of plotters to carry tools, including reciprocating cloth-cutting knives, and, in very heavy-duty versions, oxyacetylene cutting torches.

A very old flat-bed plotter is the remote handwriting machine. At the transmitter, a handheld stylus is connected to two different potentiometers which measure the stylus position in two coordinates. At the receiver, two drives move a pen stylus over paper to follow the motion of the transmitter. The coordinate system is special, the result of the coupling linkage geometry and the non-linear potentiometers, and the drives resemble large d'Arsonval galvanometers.

A second class of plotter moves the paper in one axis and a stylus in the cross axis, rather like a reversible strip-chart recorder. The stylus may be an ink pen or an ink jet. There may be an automatic stylus changer to change color and line width. Both motions are servo driven under computer command.

A third class of plotter moves the paper in one axis and has a stationary row of closely spaced styluses along the cross axis. Each stylus is pulsed electrically, one pulse per dot to be marked. A pulse may apply electric voltage to electro-sensitive paper or momentary heating to thermally sensitive paper. (Since resolution may be up to 400 dots per inch, it is a design challenge to make and wire the closely spaced styluses.)

11.6.3 Facsimile (FAX)

Most facsimile machines, FAX, are in the third class of above plotters and are operated over telephone lines. FAX transmitters scan the original optically, with closely spaced scan lines. Plain-paper FAX machines have now appeared.

11.6.4 Laser Printer/Plotter

A fourth class of plotter is xerographic with a laser selectively discharging portions of each line of the image from the electro-static charge on the selenium surface. Color is produced by repeated printing with different color toners.

11.7 AREA PRINTERS

Area printers mark an entire area simultaneously rather than sequentially, point by point or line by line.

11.7.1 Contact Printing

The first class is contact printing with light. Blueprint (diazo) machines and photograph contact printers selectively transmit, block, or attenuate light passing through a source sheet onto a print sheet which is photochemically sensitive. (The original blueprint process is obsolete but the word remains for contact printing of line drawings, including by the diazo process, regardless of color.)

11.7.2 Optical Image Printing

The second class is optical printing in which an image of the original is projected with lenses onto a light-sensitive surface. The image may be smaller, equal to, or larger than the original. The image may be anything from a line drawing to a color photograph. In photography and photo printing, the light-sensitive surface is photographic paper or film.

11.7.3 Xerography

In xerography, the light-sensitive surface is a photoelectric drum. This subject is described in greater detail in Section 2.11.2, "Lithography."

11.8 TELEPRINTERS AND MEMORIES

We have already considered electrical telegraphy as a communication medium. Early telegraphs used a human to encode the text characters into electrical Morse code pulses with an EM switch, the *telegraph key,* and a human to decode the characters after an EM *sounder* converted the electrical pulses into audible clicks. (Edison was a professional telegraph operator before he became an inventor in the field.)

Telegraphy by machine transmitter and receiver replaced Morse code with a binary code using five bits per character, *Baudot code.* EMDs converted text characters from a manual keyboard into timed on-off binary signals and decoded the binary signals into typewritten characters, the combination machine being the *teleprinter,* now called the *Telex.*

Baudot code could also be recorded on a sort of strip-chart recorder in which the strip chart, or *tape,* was sprocketed, one hole per character, and the binary code was a crosswise row of hole positions, each punched or not punched. Punched tape provided a buffer storage between slow typing and fast transmission and between reception and retransmission in telegraph offices.

Part of the evolution of the modern computer is the growth from five-bit telegraph code to eight-bit computer bytes having a larger alphabet and a parity check bit, and from punched paper tape, five bits wide, to magnetic tape, eight bits wide.

Both punched paper tape and magnetic tape, with their transports, are EM systems. Tape holes were punched by miniature EM punch presses and were sensed by mechanical pins operating contacts, by photoelectric cells, and by compressed-air and thermocouple devices.

Reaching back into time, the original binary memory machine was, and still is, the Jacquard loom. It used a sort of punched tape in the form of punched cards hinged end to end. This memory was not a code; each hole position represented a crossover point between a colored warp thread and a colored weft thread. If there was a hole, warp covered weft and vice versa. Jacquard fabrics are still a staple product of the textile industry.

Punched tabulating cards, *tab cards,* were originally invented by Hollerith in the U.S. Census Bureau to automate the analysis of census data. By hindsight, they are a variant of punched tape. Then came several generations of business machines using EM punching, handling, reading of tab cards, and computing with their data. There followed the replacement of marking with punched hole by marking with magnetized bit, with EM moving of the magnetic tape, and with electronic computing.

Read-only memories (ROMs) now include laser-marked and laser-read EM discs (CD-ROMs) of enormous capacity, adapted from the audio compact disk (CD) industry.

Marked EM media memory may ultimately be replaced entirely by solid-state memory, but as long as data must be understood by humans, there will have to be marked paper sensed by the human eye.

CHAPTER 12

Sound

You might review Section 2.1, "Sound," for a survey of sound theory before continuing with this chapter.

12.1 USES OF SOUND

EMDs are used with sound for the following functions. Some of the devices described are non-EM and are included for completeness, but it will be obvious which they are.

1. Signaling with simple signals
2. Communication (voice, music, pictures, including entertainment)
3. Probing the invisible
4. Fabrication and destruction
5. Recording and playback
6. Data storage
7. Measurement

12.1.1 Signaling

Sound *signaling,* as meant here, has a very low bit rate, often conveying only one or a few bits of information in a message. For example, a bell is used for signaling. Sound *communication,* as meant here, means conveying a bit rate suitable for speech or music. For example, a telephone is used for communication.

Signaling devices which use sound include:

- Bell
- Buzzer
- Chime
- Horn
- Siren
- Drum
- Bugle
- Gun

Most of the EM devices for signaling use a motor or solenoid to cause a hammer to impact a vibratory member, such as a bell, which is acoustically coupled to air or water. For example, there is an electric horn in which a hammer repeatedly strikes a diaphragm that is coupled to the air via an acoustic horn; the hammer is pulled against a spring by an AC electro-magnet and then released. A 60 Hz electro-magnet pulls and releases the hammer 120 times per second. Harmonics in the diaphragm yield sound with several times the fundamental frequency of 120 Hz.

A loudspeaker with an AC voltage supply is a horn.

A siren is a motor-driven centrifugal blower and flow-chopping valve on the same shaft. The flow chopper interrupts the flow of centrifuged air back to the atmosphere at whatever frequency the designer chooses. Conversion efficiency from electric power to acoustical power is quite high.

12.1.2 Communication

See Section 1.14 on communication.

Sound communication, beyond shouting range, requires an EM *microphone* to convert sound vibration to electric oscillation and an EM *loudspeaker* or *earphone* to convert electric oscillation back to sound vibration. Between that "mike" and speaker lies much of the art of electronics. Modern video communication uses no EMDs except for incidental functions such as moving lenses. (This chapter is about sound, but with the advent of TV, it is appropriate to mention video in a discussion of audio.)

(We will postulate that a device in which a supply of electrons is electro-thermally generated, electro-statically accelerated through space, and their ballistic trajectory electro-statically and electro-magnetically steered to a position on a fluorescent screen is not electro-mechanical because the only moving parts are electrons. Likewise, other vacuum and plasma tubes, mass spectrographs, and ion rockets.)

At one time, video cameras used a spinning *Nipkow disc* to scan the scene. The disc carried a spiral array of pinholes in the image plane in front of a photoelectric cell. Each pinhole scanned one image line before the next pinhole started the next image line. A synchronized disk at the receiver recreated the image from a variable brightness lamp.

An early version of color TV used an EM spinning disc containing three color filters to mix three successive images from a black-and-white CRT into a colored image. Another, synchronous, disk generated the three black-and-white images from the colored object field.

12.2 SOUND EMDs

12.2.1 Microphones

Microphones are made for frequencies from subsonic to ultrasonic. They are made to couple to the earth (for example, seismographs, geophones), to water (for example, sonar), to air (for example, speech), and to a variety of objects from steel railroad tracks (crack detection) to the human body (non-invasive examinations). Some are used as both sound sensors and as sound generators. The following principles are used:

1. In most cases, a diaphragm is used to couple the sound vibration to the transducer element. In some cases, such as in ultrasonic systems, the coupling is direct and no diaphragm is used.

2. Loose carbon granules between carbon plates change the electrical resistance between the plates depending on the pressure applied to the granules. One plate is fixed and the other is attached to the diaphragm.

3. The diaphragm is placed near a fixed plate, the two comprising a capacitor. The capacitance is modulated by the varying distance between them and is measured by a high-frequency circuit.

4. A piezoelectric crystal receives the diaphragm force and transduces it into a voltage.

5. A voice coil, Fig. 6-1, is driven by the diaphragm, and a voltage is induced in it. The same voice coil can convert motion into voltage and voltage into motion. Two voice coils with diaphragms, connected together, comprise a sound-powered telephone requiring no outside power source.

6. The earliest recording microphone was a diaphragm directly coupled to a needle which scratched a wavy line on a wax cylinder; no electricity. In a second pass over that wavy line, the waves drove the needle which drove the diaphragm as a loudspeaker. This, Edison's first phonograph, was not high-fi, but it was the first recording and playback of sound in human history. Few inventions have so little antecedent.

7. Any other pressure transducer can be coupled to a diaphragm and used as a microphone [Chapter 7].

8. Stethoscopes and speaking tubes pick up and transmit sound directly, without electricity. Two diaphragms coupled by a taut string do the same, but not very well.

12.2.2 Loudspeakers

Loudspeakers are actuators [Chapter 6] which oscillate air (and sometimes water or solids). Coupling to air is via a diaphragm and acoustic horn, or directly by a large diaphragm (usually conical for stiffness), or directly by a large, flexible-sheet diaphragm which is driven over its entire surface by electro-static force. Loudspeakers include earphones, in this discussion, which differ from other loudspeakers only in size, power, and their acoustic coupling to the listener's ear.

(This book deals only with the EM principles of sound devices; the entire subject of high fidelity, including multichannel, room and auditorium acoustics, device fidelity and distortion, and human hearing deal with subtleties which are beyond its scope.)

Loudspeakers use the following EM effects to convert electrical vibration into mechanical vibration:

1. Voice coils. These have been extensively discussed already. They are made in a tremendous range of sizes and are useful over the entire acoustic frequency band.

2. Piezoelectrics. Piezoelectrics may be used to drive diaphragms or may be directly coupled to the acoustic load. Examples of such direct coupling are sonar in water, sonar proximity-sensing in air, medical diagnostics, flaw detection in parts and assemblies, and heating in manufacturing.

3. Electro-static driving of a flexible sheet by a rigid sheet was mentioned above.

4. One of the earliest loudspeakers used an iron-core polarized electromagnet [6.3.3]. It gave way to the voice coil which introduced less distortion.

5. The loudest loudspeaker is a jet of compressed air modulated by a proportional EM valve driven by a voice coil. The aesthetic quality is not very good, but it is loud!

12.2.3 Recorders

Sound is recorded and played back on a number of analog devices and, now, on a number of digital devices. Among these are:

1. Groove and needle. Grooves were cut or molded in wax or plastic. Stereo record grooves were modulated in two independent modes, and the needle transducer moved independently in those two modes, both for recording and playing back. The original body was a wax cylinder; it was used for many years in dictating machines. Flexible plastic endless belts and flat discs were also used in dictating machines to reduce the storage bulk of the cylinders.

Recording heads always had straight-line radial motion of the head as it moved across the recording spiral or helix. Consumer record players always used a swinging arm to carry the reading head because it was cheaper, despite some distortion which resulted. In the 1950s Jacob Rabinow invented and manufactured inexpensive straight-line mechanisms for consumer reading heads and the best record players now use them.

2. Magnetic wire recorders. These were the earliest sound recorders and were later used as digital recorders in early digital computers. The wire ran between two spools.

3. Magnetic tape recorders. For sound fidelity the non-linearity of the iron oxide recording medium is eliminated by an AC bias signal. Consumer tapes carry from one to eight channels. Commercial recorders for sounds which will then be mixed carry up to 20 channels on wide tape.

Initially, the tape was hand-threaded over capstans, guides, and heads. Now a variety of cassettes are used with automatic EM engagement of the tape with the drive and head elements. The most elaborate automatic mechanism with the easiest human manipulation is the video cassette recorder, *VCR,* in which the recording track is a series of diagonal transverse lines produced and read by magnetic heads on a rotating turret.

4. Compact disk. The medium with the highest storage capacity and greatest fidelity is the *compact disk* (CD) recorder in which both recording and reading are done with a coded stream of digital dots written and read with laser light.

5. Floppy disks. Although this chapter deals with sound, recording both sound and digital data is done magnetically on similar tapes and optically on compact disks. In addition, magnetic recording of data, but not of sound, is done on *floppy disks* of the same material as magnetic tape and on *hard disks* in which the flexible tape is replaced by a rigid disk and the magnetic head floats on an air film hydrodynamic bearing [Chapter 10].

12.2.4 Other Acoustic Devices

1. Doppler flowmeter. The principles were described in Section 2.1. Doppler flowmeter sound can be introduced to a fluid conduit through the wall of the conduit. This prevents leaks in high pressure and dangerous fluid systems, such as in nuclear power plants. It also permits non-invasive measurement of blood flow in people and animals.

2. Lithofracture. Radiation from a source at one focus of an ellipsoidal reflector is gathered at the other focus. A human with gallstones is immersed in water with the gallstones at one focus of a reflector and a powerful source of ultrasound at the other focus. The sound power breaks up the stones.

3. Linear distance measurement. An ultrasonic pulse is sent along a steel rod and sensed by a movable transducer. The transit time is measured and is the measure of distance from transmitter to transducer.

4. Proximity sensor. Ultrasonic pulses are sent into the air. Any nearby object is sensed by the presence of a reflected signal.

5. Sonar. Sound is radiated into the sea. Reflected sound from submarines, fish, or the bottom is detected.

6. Medical sonar. Ultrasound is radiated into the body from a piezoelectric actuator ("transducer"), and reflections are sensed by that same transducer. Images can be formed.

7. Ultrasonic heating. Thermoplastics are heated to their softening temperature by ultrasonic power generated by a piezoelectric and focused by a solid metal acoustic horn. The heat is used either to weld the plastic or to soften it to receive a metal insert.

8. Ultrasonic machining. An ultrasonic horn, as in 7, terminates in a formed tool. The tool is lightly pressed against a workpiece with a slurry of abrasive particles between. Erosion occurs in the areas where the slurry is pressed between lands in the tool and the workpiece.

9. Musical instruments. Many EM devices have been made to play conventional musical instruments such as the violin and the piano; musical instrument museums exhibit them. Most have produced musical notes, but none have been successful because they do not produce the subtle variations of sound produced by a human musician in physical contact with the instrument. However, the pipe and electric organs are powered machines, whose only human inputs are the selection and timing of the pipes and electrical parameters.

The player piano is actually pneumatic—the holes in the piano roll are sensed by vacuum which operates the key actions via small bellows, but all notes have the same loudness. (The piano roll was one of the precursors of numerically controlled machine tools [Chapter 13].) A machine was once built which recorded the loudness and the duration of the pianist's touch. There exists a conventional recording of the machine's reproduction of a piece played by Liszt, himself, but the machine was not commercially successful.

The artistic advantage of such machines was that the tone was the undistorted tone of a real instrument; but modern high fidelity recording has made such a claim moot.

Synthetic music and speech are now created by combining electrical waveforms and playing the results through loudspeakers. This art is now developed and taught in prestigious schools of music. The electric organ was the first embodiment of this technology and musical art.

Merry-go-rounds continue to use mechanical music or recordings of it. Nostalgia rules.

The modern pipe organ's valves are EM operated by solenoids controlled by keyboard switch contacts. The keyboard contacts of both the pipe organ and the electronic organ are made without snap action and are bounce free. (In old organs, the valves were operated by the keys through mechanical linkages, some quite long.)

The steam calliope is an organ using steam instead of compressed air.

10. EM ultrasonic generators are sold to drive away rodents. Totally non-EM is the ultrasonic dog whistle.

CHAPTER 13

Manufacturing

More kinds of machinery are used in manufacturing than in any other human activity. Most of them are powered by EM motors and controlled by the spectrum of EMDs described in these pages, so in a sense, the machines are all EMDs. However, the intent of this book is to describe the engineering of devices which are themselves EMDs, not users of EMDs as components. Therefore, with the exception of robots, which bridge the definition gap, this chapter will be limited to manufacturing processes which are EM in their nature.

13.1 ROBOTS

13.1.1 History

The word *Robot* was coined by the Czech playwright Karel Capek in 1923 for his science fiction play *R.U.R.—Rossum's Universal Robots.* The play was based on the Czech word for work, and his robots were artificial people who did the work of the real people. The dream of artificial people as workers became immensely popular in science fiction, and the word *robot* is the same in all languages today.

Meanwhile real-world automation steadily developed.

After World War II, two very good engineers decided that real-world technology was advanced enough to make a first stab at a real robot. They succeeded with the "Unimate[1] Robot," which was a fixed machine which moved a single tool in a programmed manner. It was commercially suc-

---

[1] Unimate was the trademark of Unimation, Inc. and is now owned by Westinghouse.

cessful, particularly in replacing humans in spot welding auto bodies and in unloading die-casting machines.

The race was on; companies were formed; money flowed in from customers who '. . . didn't understand robots but bought one or two to see what they could do for us . . . ' At approximately \$100,000 each, robots were a prestige buy for company executives, and experimental purchases were the bulk of the sales during the heyday.

Ultimately the bubble burst in disillusionment over exaggerated claims, and the industry settled down to a small size making a small number of quite useful machines. The "artificial people robots" never materialized, although some computer people keep talking about artificial intelligence.

13.1.2 Robot Anthropomorphism

Directly emulating the human body and mind with a machine, *anthropomorphism,* is one of the themes of science fiction, but the attempts to make working hardware have resulted in huge waste in robot and computer R&D and application efforts. The continued small market and large R&D funding for these machines testifies to the passion of many for science fiction. There once was a group that had a grant to develop a robot which was to walk like a horse. There are still bursts of publicity about artificial hands and ambiguities about "virtual reality" and real reality.

There is a class of robots which approximates anthropomorphism. Each has a chain of links coupled by rotary joints in grossly simplified imitation of the geometry and motions of the human arm; "pure" versions have no linear slides at all. In fact, most of them have at least one linear slide, sometimes by being mounted on a carriage on a linear track.

On the other hand, *doing the work* of human bodies and brains with machines, designed as machines, is the historic achievement of the industrial and computer revolutions.

The pernicious effect of the science fiction image was in EM design. All animal motion is by rotary joints; there are no slides in nature. So artificial people must be made with rotary joints. The result was cantilever hinged to cantilever hinged to cantilever, all given anatomical names: shoulder, arm, wrist, finger. The assemblies had maximum flexibility and oscillation, minimum accuracy, minimum working volume, maximum cost for servo drives, and maximum cost for coordinate transformation and programming.

A human ball joint like your shoulder can rotate around any axis. But your shoulder has a complex of muscles to drive it, backed by a complex of internal position-sensing nerves and a complex of external sensors (including touch, eyes, and ears), and is directed by a wonderful controller in your head. Not easy to emulate. A complete human body has over 600 muscles and tens of thousands of inbound sensor and outbound command paths. Conversely, even 270° of rotation is not possible in human joints, as it is in simple machines, and neither humans nor any other animal has a linear slide at all.

In robot promotion, it was common to imply that the robot had a "hand" similar to a human hand. What was said above about the human shoulder is true many times over about the many joints of the human hand, to say nothing about the team of two hands used in many tasks. In all real working robots, the "end effector" is a tool designed to match the workpiece and the task, not an artificial hand.

Gradually, the humanoid vision faded and machines with conventional Cartesian motions appeared (X, Y, Z, roll, pitch, yaw). MOBOT Corporation was the first American company to make Cartesian robots and was the first purely robotic company to go public. For example, one of its robots carried 300-pound fixtures along a 400-foot row of precision NC lathes and loaded them into closely fitting chucks.

A more detailed discussion of robots is given in [A8] and there is an extensive literature, including manufacturer's catalogs.

The word *robot* has caught on and is a glamour word. It is applied to anything with an element of automation or remote control; you can get robot coffee makers.

It is useful to compare the capabilities of robots and of humans. As you read, you can replace *robot* with *machine*.

1. Robots can be made with the ability to carry any weight, any size, any distance, at any speed. Humans are limited to a narrow range in all these.

2. Robots do not fatigue, can work three shifts, seven days a week, and do not join unions.

3. Robot motions are more accurate and more uniform than human motions unaided by gauges and tools.

4. Robots need much less safety guarding than humans, and even if there is an accident, the robot cost is only money, not flesh and blood.

5. A robot requires a large capital investment; humans require only a little training to do a job no more complex than a robot can handle. (Excluded is the birth-to-hire capital cost of raising a human.) A human can be laid off if the work requirement stops; a robot cannot.

6. The installation and start-up time for a robot is long and costly; it is negligible for a human.

7. A human recognizes unspecified changes: a faulty part, a broken tool, a strange noise, and so forth. A robot must have specific sensors and programming for each foreseen fault condition. A robot may contribute to a wreck; a human will try to prevent it, or at least to limit it.

8. A human has dexterity unimaginable in a real robot. The value appears in certain machine-loading tasks and in assembly tasks. The human's hands are not very accurate, but they can use tools and instruments far beyond the scope of a robot unless the "robot" is a special-purpose machine.

9. A human can do a complex of tasks within an assignment: fill out paper, go to the stock room, deliver work, carry messages, clean up, and so forth.

10. A sick robot halts a process and requires a skilled maintainer. A sick human is instantly replaced and goes to a doctor.

11. A human is easily retrained to move from task to task. A robot can only be reprogrammed for small variations within a task. It is a major capital project to move, reinstall, retool, and reprogram a robot to convert it from task to task and to adapt the new task to the limitations of the robot.

13.1.3 Industrial Robots

Industrial robots are *real* machines which grip and move either workpieces or tools. Practical robots are machines having combinations of linear slides and rotary joints. Both are referred to as *axes*. In general, the linear slides provide displacement and the rotary joints provide orientation. Motion envelope dimensions vary from a few inches to hundreds of feet. Axis actuators may be pneumatic, hydraulic, or electric.

Point-to-point robots transfer workpieces or tools from one defined place to another. *Continuous-path* robots move a tool such as an arc-welding electrode or a paint spray gun along a defined, smooth, curved path.

For example, point-to-point *workpiece* transfer tasks include loading and unloading machine tools and loading components into printed circuit boards *(stuffing);* point-to-point *tool* transfer tasks include moving spot welders to many points on auto bodies. Continuous-path tasks include moving an arc-welding electrode or spray gun.

Programming the positions of a point-to-point robot is done by entering the coordinates of its successive positions into its electronic controller.

Programming the path of a continuous-path robot is usually done by switching it to *teach mode* and moving transducer inputs near the gripper to control the axis actuators.

In all cases, special tooling is required to enable the robot to grip the part or tool. Such tooling is called *end effectors.* The end effector may be a simple gripper such as a pair of jaws or a vacuum cup, it may contain mechanical self-aligning mean *(passive homing guidance),* and there may be position sensors feeding back to the robot's drives *(active homing guidance).* Homing guidance compensates for random variations (tolerances) in the positions of the parts, tools, associated machines, and the robot itself.

Some robots are *branched:* They have more than one motion generator moving more than one end effector. This configuration is particularly useful in loading and unloading a machine tool: one end effector removes a finished part and the other immediately loads a new part. There is no wait for the first to deposit the finished part and then get the new part.

Robots have logic connections to their associated machinery for starting, interlocking, and so forth. The robot controller itself is electronic, typically a PLC (Programmable Logic Controller) for logic and a numerical control system (see below) for continuous-path definition.

13.1.4 Economics of Robots

A few words on economics will help put robots into perspective. (Some of this section recapitulates Section 13.1.2.)

With few exceptions, robots merely replace people who are capable of doing the same work. Furthermore, people can be quickly hired and fired, or at least transferred, as needed while robots require a long lead time to buy, install, and debug, and require a capital investment which cannot be recaptured if production slows.

People can be reprogrammed for major changes (for example, from operating a lathe to operating a milling machine); robots can be easily reprogrammed for *variation within task* but are expensive to modify for *conversion from task to task.* Furthermore, humans will reject faulty parts, fill out paper, and have far more dexterity than any robot. So after the romance has faded, why bother? (The buzz word in industry is *justify.*) This is why:

An installed robot costs roughly the same as a human worker, including overhead, for 1 year. The usual justification for buying a piece of capital equipment is that it pay for itself within 1 year. So, a one-for-one change has no great appeal to management. *But,* if the factory works two shifts, the investment is highly profitable and has great appeal. Three shifts is heaven to the robot salesperson.

Next, the equivalence of one robot to one human is a loose generalization. One factory task may be moving a small, 1-ounce part 1 inch, but another task may be moving a large, 50-pound part 20 feet. Humans come in a very narrow range of sizes. The range in repetitive lifting capacity between a small woman and a large man is only about five to one and the range in reach is only about two to one. The range in lifting capacity and reach of available robots is many orders of magnitude. *Therefore, the productivity advantage of robot over human is much greater when the task involves big size, heavy weight, and long distance.*

Next, safety. The cost of a sprained back is roughly the cost of a robot. And there is no assurance that there will not be another sprained back next week. Furthermore, dangerous machines tended by humans must be expensively guarded, and the guarding slows down the work. There are many horror stories of humans (workers, not managers) who cleverly evaded the guarding to increase their own output or convenience and left hands behind. If a robot end effector gets smashed, it's only money lost. And there is no slowing due to sliding safety doors.

Most of the above justification is based on non-humane, cold, money. Much has been written about robots improving the quality of life by relieving humans of dreary labor, but there are few factory managers who would spend cold money to improve "quality of life" unless there were also clear benefits in money. However, many managers prefer to have unemotional and non-unionized machinery work for them instead of temperamental humans.

This has not been an exhaustive analysis of robots in the real world by a wide measure, but it should give you a glimpse.

13.2 REMOTE CONTROLS

It is useful for a human to control a machine from a distance. The distance may be only a few feet (as at a radioactive hot cell), or may be several miles (as in the case of a remote power plant), or may be thousands or millions of miles (as in the case of a satellite or space probe).

A remote control system is usually one or more closed-loop servos with human command inputs and with both forward and return communication paths. The paths may be wire, or radio, or fiber optic, or IR, or visible light, or ultrasonic (to submersibles). Commonly, the controlled machine is partly automatic to minimize the number of control channels and, in the case of space distances, to compensate for transmission time. (But a remotely controlled detonator needs no feedback path.)

Many people think of remote control as a branch of robotics, as in the "robot arm of the space shuttle," although the whole point of robots is to automatically perform a programmed cycle and eliminate a human. (The programmed cycle may not be fixed; it may vary under the control of a hierarchically senior computer and a variety of sensors, and it may also be under remote control command.)

Many remotely controlled manipulators, such as on the outside of submarines, or the space shuttle, or in hot cells, or "cherry picker" baskets for carrying human workers, are made as jointed arms with controls operated by humans who watch them move.

13.3 ANALOG CONTROLS

EMDs are used in manufacturing for controlling position and motion, temperature, chemical composition, and other parameters. Older systems are analog in nature, and while, more and more, digital systems have taken over, there are many places where analog technology is still simpler and cheaper. This section will survey some of the analog systems in use, and the next section will describe digital systems.

13.3.1 Machine Tool Tracers

Some machine tools make duplicates of handmade models. For example, rifle stocks, which are complex three-dimensional shapes without geometrical definition, are made on special lathes which duplicate the shape of a model. Typically, motion along one axis is at a uniform rate while motion along the cross axes is controlled by servos whose inputs are transducers which sense the surfaces of the model.

Some tracer-controlled machining duplicates the shape of the model, while other tracer machining makes a negative of that shape; the negative becomes a molding die.

Tracer systems are made both hydraulic and EM. An interesting EM transducer uses a short, low-current, electric arc between the model and an electrode. The voltage drop across a constant-current arc varies with the arc length and is the transducer output. There is a demonstration in which a huge die-making machine follows the shape of a piece of crumpled aluminum foil without deforming the foil.

Photoelectric tracers follow a line on paper or the edge of a piece of paper cut to a shape. Many flame-cutting machines use such photoelectric tracers. It is economical to flame cut even a single piece using a drafter's drawing or cutout as a *template*.

Cam-operated machines such as automatic screw machines (small lathes) are analog-controlled machines without amplifiers between the cams, which are shaped to define the motion, and the machine slides. A set of cams on a common shaft directly drives the slides.

13.3.2 Temperature

A furnace temperature which is programmed as a function of time may use a paper template turned by a clock motor and followed by a sensitive stylus as a program controller.

Temperature transducers feed electronic controllers having stabilizing circuits. The controllers feed actuators which may be electric power contactors or electronic amplifiers or fuel valve actuators.

For many years, all-pneumatic systems comprising transducer, amplifier, and actuator controlled fuel-fired furnaces. Now, combination EM and pneumatic systems are also used. It is still difficult to make an EM throttling valve operator which can compete with a pneumatic diaphragm operator in cost and in reliability under severe corrosion conditions.

13.3.3 Chemical Composition

Chemical composition applies, for example, to steel, paper, petroleum products, paint, and drugs. Control of fluids is by valves, usually throttling (as for fuel in temperature control). Control of solids is either by batch weighing or by continuous weighing on a moving conveyor. EM transducers with electronic amplifiers and EM actuators are extensively used, but with pneumatic competitors.

13.4 DIGITAL CONTROLS

13.4.1 Numerical Control of Machine Tools

In 1950, at MIT, a milling machine was demonstrated in which each axis was servo driven. The servo inputs were streams of electrical bits, derived from a punched-tape memory, and each servo served as a digital-to-analog converter. Thus began a gradual revolution in the programmed control of machine tools.

A word about the word: Digital computers were only a few years old so *digital* was not yet a common word. *Number* was, so the system was called *numerical control,* NC, and its descendants still are.

Variations were spawned. There is *point-to-point NC* in which an EM *absolute encoder* is the feedback transducer and is not subject to start-up or power interruption errors; and there is *continuous-path NC* in which an EM *incremental encoder* feeds back a stream of bits for comparison with the command stream.

Source data came first from punched tape, then from magnetic tape, sometimes over wire from a central controller. Computer programs were developed to convert drawing dimensions to command data; the first use of voice-responsive computers was for humans reading drawing data into these computers. Now computers are often incorporated into the machine tool controllers, the technology is *computer numerical control,* or *CNC,* and the programming is done by a trained machinist.

Most NC machine axes are driven by electric motors, although hydraulic servos can also be used.

13.5 EM WELDERS AND HEATERS

Arc welders use the heat of an electric arc to melt the metal being welded. Plasma arcs are basically electric arcs forced outward by a gas jet.

Spot welders use I^2R heating of two workpieces clamped, under EM control, between a pair of electrodes. Seam welding is spot welding in which the electrodes are wheels and a rapid succession of electric pulses produces a succession of closely spaced spot welds.

Plastic sheets are welded either by heat conducted in from hot platens or by dielectric heat induced by RF power through electrodes, which also serve as clamping bars.

Plastics are welded, or merely heated for forming, by ultrasonic power from a solid metal acoustic horn driven by an ultrasonic actuator.

Metal parts, mostly steel, are induction heated by an RF magnetic field generated by a coil into which they are inserted.

Laser cutters and welders use heat from laser radiation from electrically powered lasers.

Electron beam welders use high-power cathode rays in a vacuum chamber. The electron beam generator is usually stationary and a remote control or programmed mechanism moves the workpiece under the beam.

Small electric arcs between a tool and a workpiece are used to erode the workpiece in *electric discharge machining*, EDM. The tool may have a shape to be impressed on the workpiece or it may be a fine wire under tension which acts like a band saw. A tap disintegrator uses electric sparks to disintegrate a tap broken off in a hole.

13.6 INSPECTION DEVICES

EM inspection instruments and machines are pervasive in industry. The EM portions may be measuring transducers, workpiece manipulators, or both. Among such machines are an automatic EM machine to gauge screw threads, another to manipulate artillery shells through a gamma-ray beam to detect voids in the explosive charge, and another to carry a radioactive source through a large pipeline to X-ray its welds [7.4.3]. A large inspection machine uses a neutron beam from a small nuclear reactor through which an EM machine manipulates airplane wings to detect internal corrosion.

13.7 MAGNETOSTRICTIVE DRIVE

Magnetostriction has been used as a driver of sonar transmitters, competing with piezoelectric drivers.

There has been a commercially successful magnetostriction actuator for moving the grinding head of centerless grinders [6.18].

13.8 MAGNETIC REPULSION FORMING

When a DC pulse is applied to a primary coil near a solid conductor, the rising magnetic flux induces a secondary current in the solid conductor, which is a single short-circuited turn. The direction of the secondary current is such as to oppose the rise of the flux. The two currents react with the flux to repel each other.

If the short-circuited solid conductor is a ring coaxial with the primary coil, the repelling force tends to shrink the ring. If the primary pulse is extremely large, such as from the discharge of a large, high-voltage capacitor, the force can be large enough to deform the secondary solid conductor inelastically and shrink it over a third body within it. This, in effect, is electrical swaging. Auxiliary secondary rings can be used to shape the field. This process is commercial [6.16].

And, of course, there are innumerable motor-driven hand tools.

CHAPTER 14

Assembling and Connecting

14.1 GASES, LIQUIDS, SOLIDS, AND ELECTRICITY

14.1.1 Conduits and Valves

Liquids, gases, and electricity are all fluids; they take the shape of their containers and they flow along conduit portions of those containers. Liquids and gases are not restrained by air and will spill out if not confined by enclosures with solid walls and well-sealed connections; electricity flow *is* obstructed by air, most liquids, and most solids. It will *not* spill out of its conduit (except under very high voltage), and it requires only touching of conduits to pass from one to another. These properties are among the reasons that electricity is the fluid of choice for transmitting power and information.

Electricity conduits are *conductors,* flow obstructors are *insulators,* intermediate materials which permit some flow are *resistors,* and means to establish and to interrupt touching are *connectors* and *contact makers.* The variety of conductors and connectors is a principal subject of this chapter.

(Transistors and vacuum tubes are electrically controlled variable resistors. Transistors and thyratrons which can switch from zero to infinite resistance compete with the contact makers of Chapter 9. Those which switch on only a portion of each AC cycle as a form of pulse-width modulation are, in effect, lossless resistors. Their switching speed makes it impossible for contact-making EMDs or EM valves to compete in this mode. This is not an attempt to summarize electronics in a paragraph; this book is intended to cover only electro-mechanics, with some comparisons with other technologies.)

We start and stop the flow of liquids and gases with valves which change the shape of the enclosing conduits; we start and stop the flow of electricity merely by making its conduits touch and separate (and by the electronic means of the previous paragraph).

Inserting an air gap is equivalent to closing a valve. The regions where conductors touch and separate are *contacts*. There are very many devices in which contacts touch and separate and others in which they touch and slide. Contact technology is the subject of Chapter 9.

14.1.2 Power Distribution

Power within factories was once distributed by long rotating *line shafts* tapped off to individual machines by belts and pulleys and switched on and off by manually operated clutches. At first, such power came from a steam engine or waterwheel, then from a single large electric motor. The individual electric motor for each machine, drawing power from a distribution network of electrical conductors, was a twentieth century development.

Pneumatic power, both within factories and in the field, is still widely used, primarily for linear actuators and certain hand tools, but also for certain transducer and information transmission functions. Pneumatic power will probably remain in coexistence with electric power. Pneumatic power in factory and field is generated by a central compressor. It is distributed by a pipe and flexible tube network to individual machines, tools, and other devices. In the factory, the compressor is usually driven by an electric motor; in the field, usually by an engine.

Hydraulic power is always generated near the point of use because of the power losses in pipe transmission; this is a major impediment to its wider use. Hydraulics usually wins where a high force-to-weight ratio is valuable in actuators and where a high-performance linear motion servo is desired.

Pneumatic tools compete with electric tools. Pneumatic and hydraulic linear actuators for limited travel between two fixed positions usually win over electric actuators.

Manual, pneumatic, and EM valves are used in pneumatic systems, and manual, hydraulic, and EM valves are used in hydraulic systems.

14.2 CONDUCTORS

14.2.1 Materials

Most electrical conductors are copper or aluminum, but the following materials are also used:

1. Gold wire is used to connect semiconductor chips.

2. Gold is electro-plated onto electrical connector contacts to provide tarnish-free surfaces.

3. Gold, silver, palladium, rhodium, and platinum are used in electrical switching contacts, usually as alloys. Chapter 9 lists most contact alloys.

4. Pure iron rails are used in railroads as *third rail* conductors on which contact shoes slide. The steel railroad rails on which the wheels roll also serve as conductors.

5. Refractory metals are used inside vacuum tubes and for electrical contacts which must resist frequent use, arcing, and welding.

6. Copper-plated steel wire is used as the central conductor in some coaxial cables and in TV antenna wire.

7. Brass is used for current-carrying mechanical parts such as wiring terminals.

8. Phosphor bronze, beryllium copper, nickel silver, and sometimes stainless steel are used in current-carrying springs.

9. High nickel alloy steel is used in the form of long helical springs as the conductors between cardiac pacemakers and the electrodes touching human hearts. This construction has the longest mechanical fatigue life available. (Beryllium copper would be better if it did not poison the patient. Stranded graphite fibers were forbidden by the FDA for reasons hard to believe.) [16.4].

10. Carbon electrodes are used in arc furnaces. Carbon contacts are used to resist arcing and welding, and carbon sliding brushes are used against commutators and slip rings in motors and generators. ("Carbon" includes graphite.)

11. Mercury is used for switching contacts in many kinds of devices.

12. Soft and hard electrical solders use lead, tin, silver, gold and other metals alloyed in various proportions. Tin is used as a plating over copper wire to make it easier to solder.

13. Zinc sacrificial electrodes deplate (corrode) to the ground or to the sea to prevent corrosion of the adjacent metal.

14. Doped silicon, germanium, and other semiconductors are used in semiconductor devices.

15. Stainless steel, iron, nichrome alloy, and carbon are used in resistors and heaters.

16. Stainless steel and carbon are used as electrodes in chemical processes in which they are subject to chemical attack.

14.2.2 Forms

Copper:

1. Solid round wire, with diameters from less than 0.001 inch to over 1.0 inch, bare or tinned

2. Stranded round wire, bare or tinned

3. Flexible braid, round and flat. Braided outer conductor of coaxial cable

4. Woven screen (for RF shielding)

5. Rectangular wire for coils. Such coils have minimum cross-sectional area reserved for insulation.

6. Round tubing of many sizes. Used as conductors with minimum skin effect and maximum mechanical stiffness and strength

7. High-voltage transmission line tubing, assembled from tongue-and-groove strips, helically laid, with the outside diameter chosen to prevent corona

8. Rectangular bus bar, for high current

9. Thin ribbon, stacked for flexible leads in high-current machines such as spot welders. Also wound into spiral coils

10. Printed circuits. Thin sheets of copper are laminated to plastic *substrates* and then etched by photolithography into intricate circuits to which components are soldered. The circuits may include connector contacts and ground planes. Several layers may be laminated together after etching. Microwave stripline is a special case of printed circuit in which the substrate is ceramic.

11. Figure-eight cross section, used as sliding bus bar. (One lobe of the eight is clamped by support fittings, and the other lobe is the track for sliding brushes.)

12. Hollow conductors carrying cooling water

13. Coaxial cable, round waveguide, and rectangular waveguide. RF and microwave radiation travel inside their hollow spaces. Electricity is conducted, by skin effect, along a thin layer of their surfaces facing the hollow space.

Aluminum:

1. Most of the same forms as copper (above)
2. Foil ribbon for shielding other conductors

3. Foil strips cut to specified length to reflect radar as a countermeasure

Other materials:

1. Thin sheets of resistance material, laminated to silicone rubber, then etched into patterns, for flexible heaters

2. Electro-plated films for combinations of corrosion protection and electrical conductivity, particularly at RF where skin effect forces current to the surface

3. Rivets and other formed solids for installation as electrical contacts

4. Springs of all forms: beryllium copper for minimum fatigue and drift, stainless steel, nickel silver, high nickel alloy steel for cardiac pacemaker leads having maximum fatigue life

5. Combination mechanical and electrical parts: links, stampings, castings, and so forth

6. Conductive films of flame-sprayed or vacuum-deposited metal are deposited inside plastic housings to shield radio frequency radiation, inward and outward

7. Microwave waveguide and cavities

8. Vacuum-deposited aluminum on thin dielectric film, to be wound into capacitors

9. Powder suspended in adhesive, for conducting inks and adhesives

14.3 CABLES

A *cable* is an assembly of two or more conductors lying side by side, usually to extend circuits over a distance. *Continuous cable* is manufactured in long lengths, wound on reels, and cut off at desired lengths. A *cable form* is assembled as a component of an electrical assembly, has a shape determined by the location of components to be interconnected, and has individual conductors entering and leaving at locations near the components they connect.

There is an immense variety of continuous cable manufactured. Some of the variations are listed here:

1. Shielded cable
2. Coaxial cable
3. Twisted pairs
4. Color-coded conductors

5. Printed number-coded conductors

6. Solid conductors

7. Stranded conductors

8. All combinations of the above, in any quantities, in substantially round bundles, helically laid for flexibility (like rope)

9. Printed circuit cables, sometimes integral with printed circuit assemblies, built on thin, flexible, insulating substrates

10. Cables molded, with rubber jackets, into helical form for flexibility and extensibility

11. Cable and connectors, prewired, and molded into a single rubber jacket

12. Hybrid cables comprising both electrical conductors, fluid-carrying tubes, and mechanical-tension wire rope

13. Flat cables of wires laid side by side and joined by a common extruded insulation or by other means [Fig. 14-2].

14. Optical fibers are included in cables and are treated in substantially the same way as metal conductors. Cables may carry both optical fibers for data transmission and copper conductors to conduct power to spaced amplifiers for the optical signals.

14.4 FLEXIBLE CABLES

Portions of many electrical assemblies must move with respect to each other, so the cables joining them must be flexible. In some cases, flexing occurs only once, when the cable is installed. In other cases, the flexing is occasional, as when a door is opened. In other cases, the flexing is cyclic, as when a machine carriage moves back and forth. Cyclic flexing may occur many millions of times during the life of a machine. Flexing may be in bending or in twist or in both.

14.4.1 "Twisted" (Helical) Cables

If you merely bundle a quantity of parallel wires they will become a stiff beam, as shown in Section 2.3. The wires on the outside of a bend are in tension, the wires on the inside of the bend are in compression, and the whole is almost as stiff as a copper bar of the same size.

But if the individual wires are in the form of a helix, then each wire starts in tension for one-half turn of the helix and in compression for the next half turn. A very small amount of endwise sliding eliminates both tensile and compressive stress, and the cable is flexible.

You could make this helix by twisting the original bundle, but if you tried to do so you would impose a twist on each individual wire and as soon as you let go the bundle would twist back to its original form. Try it with just two pieces of wire or cord a few feet long.

Next, give each wire a reverse twist as you give the cable a forward twist so the individual wires remain untwisted, just bent into their helixes. The cable will lie quietly in its helical form. Try that with the same two pieces. The individual wires are said to be *laid* into the cable. This was learned about rope several thousand years ago and rope has been laid ever since.

Tubular *braiding,* as in the outer conductor of a coaxial cable, does the same thing with alternate strands on clockwise helixes and the others on counterclockwise helixes. Flat braiding either uses a zigzag path for each strand, as in a girl's braided hair, or else is flattened tubular braid.

Common usage speaks of "twisted pairs" and "twisted cables." The words do no harm, but the wires and cables are actually laid, or "helixed," not "twisted."

14.4.2 Flat Cables

Maximum flexibility in one direction is achieved by laying the strands side by side in a ribbon. The most flexible electric cables are printed circuits with parallel conducting stripes and very thin insulator substrates, as thin as 0.001 inch. In machinery, flat cables are assembled from several electric cables (which, themselves, may each have many conductors) and from fluid hoses. They are either tied side by side to a steel ribbon, or lie inside hollow chain links as in Fig. 14-1, or are molded into flat jackets as in Fig. 14-2, Section A-A. The chain-link system may have cable clamps which hold the individual cables and hoses side by side with a common neutral axis, or may merely provide a loose jacket inside which are the unrestrained cables.

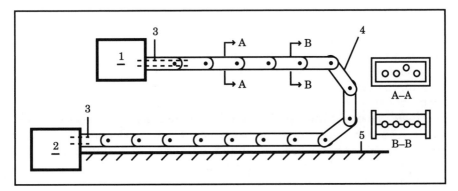

Figure 14-1   Flexible Cable Carrier

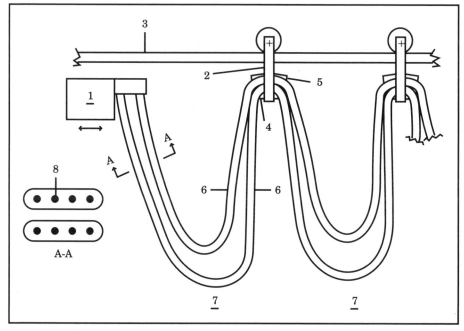

Figure 14-2   Festoon Cables

14.4.3 Flexing Modes

Flexible cables must accommodate motion between their ends. The motion may be from a fraction of an inch to over 100 feet, bending of over a revolution, and twist up to a full revolution. A cable from a fixed point to a handheld tool such as a welding torch or a telephone may accumulate many turns of both bend and twist.

We have all seen *kinked* telephone wires and garden hoses. *A kink is a small radius bend, and it breaks wires. If a cycling machine kinks its cable, it self-destructs.* Kinking usually takes place next to a cable's end anchor, where flexibility suddenly goes to zero.

Twist: It is as difficult to twist a finished cable as it is to make it by twisting, so the best way to provide twist is to provide one or more spiral turns which bend.

Bend: Helical cables bend; they are made helical so that they do so easily. The only question is the permissible radius of bend. The shorter the helix pitch and the looser the bundling, the more flexible is the cable. The limit is the flexibility of the stiffest strand. In machinery, this may be an air hose or hydraulic hose.

Extend: Extension is achieved in several ways:

1. By opening the radius of bend as in the festoon cables of Fig. 14-2. In the figure, the festoon provides flexibility without kinking and travels up to several hundred feet. Device 1 moves in a straight horizontal line. Festoon carriers 2 roll passively on parallel track 3. Each carrier has a curved saddle 4 and clamp 5 between which one or more cables 6 are clamped. The cables hang in flexible festoons 7 between the carriers. The cables are made flexible by being laid in a single plane of parallel conductors 8.

2. By shifting the bend points as in the hollow-chain flexible cable carrier of Fig. 14-1.

3. By unrolling the cable from a cable reel, as in Fig. 14-3. Figure 14-3 shows a reel and slip rings. Cable 1 is wound on spool 2 (the *reel*). Its inner end terminates on coaxial slip rings 3 whose brushes 4 are connected to fixed cable 5. Large spiral spring 6 maintains tension on cable 1 and reels it in when it can.

4. By alternately supporting the cable rigidly and providing short flexing portions. Figure 14-4 shows the accordion cable carrier system used in airport telescoping docks. The dock sections carry tubes 1 hinged in such a way that their adjacent ends remain close. The cable is carried within the tubes except for omega-shaped (Ω) portions 2 between tubes which flex to accommodate the change in the angle between tubes.

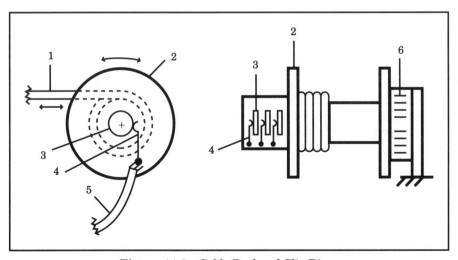

Figure 14-3   Cable Reel and Slip Rings

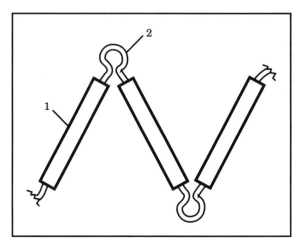

Figure 14-4   Accordian Cable Carrier

5. When the motion is vertical instead of horizontal, gravity forms the cables into a hanging U and no carriers are necessary. Elevators are connected through such hanging cables.

6. A common form of flexible "cable" is a short single conductor between a fixed point and a moving contact member on a relay, contactor, motor starter, circuit breaker, or spot welder. Typical forms include a coiled *pigtail* of finely stranded wire, a U-shaped flat braid, and, for very heavy currents, a stack of U-shaped thin strips. Flex life may be millions of cycles, so detail design is critical.

Another and very common way is to throw in a bunch of loose cable and hope for the best.

A responsible designer considers the cabling (and plumbing) of a machine as an integral part of the electro-mechanical design. A responsible designer does not delegate the engineering to electricians and plumbers with a note, "Wire and plumb to suit." If the designer does, we get engineering by electricians and plumbers, and it looks and acts like it.

14.5 PROTECTING CONDUCTORS

14.5.1 Bare Conductors

Some conductors are left bare. For example:

1. High-voltage transmission line conductors are suspended in the air at a substantial distance from any other object than their supporting insulators. This prevents arcing. No further protection is necessary.

2. Rigid bus bars in power substations and in chemical plants are left bare because their rigidity prevents them from coming close to other objects and they are guarded from people and animals by fences.

3. Electric fence wires are deliberately left bare because they are intended to shock animals that touch them. (People are expected to know better.)

4. Solid conducting portions and short flexible leads of devices such as contactors are left bare because they cannot accidentally touch other conducting bodies. People are protected by overall enclosures or sometimes by the very low voltages used.

5. Electrical and magnetic shielding comprise a class of enclosure. They prevent external electric and magnetic fields from inducing voltages in an enclosed conductor or entire enclosed device. They also isolate the outer world from fields generated by the conductor or device.

In most other cases, some enclosure of the conductors is necessary to prevent electrical conduction to other bodies, including people, and to prevent fire from an electrical fault within the enclosure. These enclosures do not appear in electrical circuit diagrams or electrical circuit calculations, but they are essential electro-mechanical portions of real equipment.

14.5.2 Insulated Conductors

Conductors are enclosed in insulation to prevent unwanted currents when the conductors touch other conductors. Examples include:

1. Enamel, a thin layer, intended to insulate adjacent turns in a winding. There is a range of temperature resistances available, up through ceramic coatings. Some enamels act as solder flux when they are heated.

2. Extruded plastic enclosing the conductor. Available in many colors and sometimes color striped for further identification. Vinyl, nylon, and Teflon are typical materials. Sometimes two concentric insulating layers are provided, the outer one having greater abrasion resistance and the inner one having better electrical properties.

3. Those cables which are manufactured in continuous lengths usually have a protective jacket. The jacket may have layers of helically wrapped cloth or paper or shielding foil, then extruded plastic—usually an elastomer—and sometimes an outer jacket of lead.

4. Cables intended for laying on the ocean floor may have an outer jacket of steel wires, also helically laid, to protect against both abrasion and *fish bite*. Oceanographic instrument cables may be enclosed in plastic containing tiny glass hollow spheres, *syntactic foam,* which make the cable neutrally buoyant despite high ambient pressures.

5. High-voltage cables may be filled with either oil or pressurized gas to displace air and increase the dielectric breakdown strength.

6. Both in machines and in buildings, it is sometimes considered that cable insulation does not offer sufficient protection from the environment and from accidental or intentional abuse. Then the cable is enclosed in solid metal pipe *conduit,* flexible conduit of helically wound sheet metal, or sheet metal rectangular duct. For machine tool and similar service, oil-tight fittings are available to connect conduit to enclosures. Similar fittings, called *terminal tubes,* are used on ships to keep out seawater. Such hardware may not interest many people but is essential for the success of equipment installed in the real world.

14.6 PROTECTING FROM CONDUCTORS

At one time, asbestos fibers were wrapped around conductors and held in place by cotton. It was an excellent electrical and heat insulator. It was then discovered that the fibers had tiny hooks which caught in the lungs when breathed and caused cancer. The law of unexpected consequences had struck a truly terrible blow. The same thing happened with PCB, a superb liquid electrical insulator and dielectric, which also turned out to be a carcinogen.

Most such tragic effects occur with chemicals whose submicroscopic effects on living tissues are difficult to predict. However there are others.

Push buttons which must be operated frequently by force parallel to the finger axis can cause *carpal tunnel syndrome* in operators.

Magnetic fields from power transformers have recently come under suspicion of being carcinogenic to children.

Of course, visible dangers and safeguards, from stepladders to door interlocks, and known sources of danger such as X-rays are constant subjects of engineering thinking.

14.7 WIRING HARDWARE

"The devil is in the details" applies to engineering design as well as to legal and political design. There is an entire industry devoted to wiring details, the *hardware,* of wiring: wire and optical fiber terminations, connectors and terminal blocks, labels, mounting and fastening, insulating, cable forming, ties and anchors, strain reliefs, mounting rails, cable duct, cable fittings, and more. And acres of specs. Unless you are engaged in wiring design these components and materials probably rank as deadly dull, so this chapter will not fill space which you would skip. Just know that they are there—you may need them some day.

14.8 ASSEMBLING AND CONNECTING

Electrical packaging is the supporting structure, enclosure, component mounting, and wiring of an electrical system. It can be as simple as a sheet metal box or as complex as a supercomputer in which heat dissipation and short conductor lengths are the ruling objectives. It includes making enclosures abuse resistant, miniature, attractive, and accessible for maintenance.

Almost all components are mounted and interconnected on flat panels. For example:

1. For power and machine tool controls, the panels are steel and are either mounted inside thin steel cabinets or inside the walls and doors of deep steel cabinets. (In olden days, panels were of slate, marble, or insulating compounds. Now we depend on small insulating members between conducting members and ground.) Interconnection is by individual conductors bundled into neat, branched cables.

2. In the days of vacuum tubes, components were mounted on the outside of aluminum boxes, *chassis;* their terminals extended through the box wall; and wiring was inside the box. Some small components were also inside the box.

3. The printed circuit board, invented and developed during and since World War II, is as fundamental to modern electronics as is the transistor. It provides mechanical support, complex electrical interconnection, dense spacing of tiny components and conductors, and zero labor per individual connection.

The integrated circuit (IC) is itself a subminiature printed circuit, made integral with its connected components, all on a single semiconductor chip.

The edge connector, often part of the printed circuit, permits circuit modularization with a board as the module and easy replacement of modules. Components are connected to the printed circuit board either by leads extending through holes and soldered to the conductors or by leads soldered to the conductors on the same side as the components, without holes, and with the components adhesively mounted. *(Surface mount.)*

Microwave stripline is printed circuitry on a ceramic substrate.

Sometimes connector sockets are soldered into the printed circuit and components, such as IC chips, are plugged into the sockets.

Printed circuit boards are plugged into a plane of connectors which themselves are either interwired or soldered into a *motherboard* which provides the interwiring.

The supporting structure of a stack of printed circuit boards provides drawer-like slides for the individual boards and aligned support for the motherboard. It is sometimes referred to as a *cage.*

So far we have ignored volume, shape, and aesthetics, a rare luxury. There is usually pressure on the packaging designer to use less volume than appears possible. The designer must often conform to an arbitrary envelope established by considerations other than cramming in the circuitry and meeting electrical specifications. The images arise of the shoehorn as the primary design tool and of the old British expression "blivet" (ten pounds of electronics in a five-pound bag). More on these subjects in Part 3, "Design."

14.9 PROTECTING FROM THE ENVIRONMENT

We have considered protecting the environment from electrical conductors; we must now consider protecting the equipment from its environment.

Some electrical equipments operate in temperature and humidity controlled, air-conditioned rooms with regulated power supplies; but some do not. Here is a list of environmental threats which must be met by the mechanical design of electrical equipment. It will be left to your imagination to realize where each of these conditions is encountered:

- Shock
- Vibration
- Heat
- Humidity
- Weather
- Corrosive liquids and gases
- Oil and other liquids
- Cleaning fluids, including steam jets
- Fungus and marine growths
- Roots
- Large animals, including fish
- Unintended negligence
- Intended sabotage and vandalism
- Nuclear radiation
- High and low external pressure

There is no room in the chapter of a general book to describe all the means taken for protection, but here is a list of some of the elements used:

- Cushions, dampers, and strong and rigid mounting for shock and vibration

- Desiccants for humidity
- Sealed enclosures for hostile fluids (liquids and gases) and pressures. There are degrees of sealing, from a drip roof to enclosures that exclude deep-ocean pressures.
- Poison coatings for fungus and marine growths
- Conformal coatings for fungus and hostile fluids
- Strong and locked enclosures for mechanical attacks
- Corrosion-resistant materials and coatings
- Protecting from negligence, incompetence, and vandalism by operators, maintainers, and passers-by is one of the great challenges to design. It really rates a long chapter by itself. Some ideas on the subject are in Part 3, "Design," of this book.

14.10 HEAT DISSIPATION

All electrical devices and conductors generate heat; the only exceptions are cold superconductors. Unless this heat is transferred to the environment, the device temperature will rise to a destructive level.

In many systems, passive convection to the atmosphere is sufficient; no special provision is required. In many others, heat dissipation requires mighty engineering effort. Among the techniques used:

1. Fan cooling within a cabinet to spread the heat over the entire cabinet surface
2. Fan cooling circulating external air through a cabinet
3. Refrigerating the air within a cabinet
4. Circulating a refrigerated coolant within tubing distributed within a cabinet
5. Use of a vortex tube to generate cooling air
6. Use of massive copper or aluminum heat sinks to conduct heat from hot spots. Use of circulating refrigerated coolant through the heat sinks.
7. Use of heat pipes
8. It is sometimes necessary to cool radiation detectors to cryogenic temperatures. Josephson junctions (reverse thermocouples) do this electrically. Liquified gases evaporating provide cryogenic temperatures.
9. It is sometimes necessary to heat a chamber for testing devices or, in production, for *burning in* devices to find faulty components. Electric heaters with control thermostats are used.

CHAPTER 15

Military Devices

Before World War I, the military were notorious for their hostility to new technology. The story is told that Elmer Sperry, the inventor of the first aircraft gyro instruments, was turned down so often in the United States that he went to England with them, succeeded, and was then accused of selling military secrets to a foreign power. Similar stories and cartoons are legion. Gradually, the culture of conservatism changed, and now the military are a principal source of R&D budgets.

This chapter is a brief survey of military EMD technology. It does not include material covered elsewhere in the book, such as servos, inertial components, and antennas.

15.1 SAFING, ARMING, AND FUSING

Safing (a contraction of safetying) is making a pyrotechnic or other device resistant to unintentional effects which might ignite or start them.

Arming is removing safing, setting up fusing, and making the pyrotechnic or device ready to respond to the fuse.

Fusing is providing means to fire the pyrotechnic under the desired conditions. A fuse is a transducer.

Elaborate arts were developed for these functions because military devices are subject to severe conditions, yet unwanted explosion of munitions must be prevented. Ignition *must not* be initiated by abuse or accidents and yet *must* be initiated by conditions which may be quite subtle. For example, an antiaircraft shell with a proximity fuse must withstand handling shock, vibration, temperature extremes, long shelf life, and the final shock of being fired from a cannon, *without* exploding, but it *must* explode in response to a faint change in its RF field when near an airplane in the air.

At one time, flame in a slow-burning pyrotechnic cord initiated pyrotechnics, and mechanical clockwork provided timed operation. A basic technology introduced in the twentieth century is the use of electricity in safing, arming, and fusing. The initiating element is a small resistor, called a *carbon bridge,* which is coated with a sensitive pyrotechnic and implanted in the main pyrotechnic, usually in duplicate for reliability. Now all the techniques of electrical control can be applied. The following are examples of electrical techniques. Some versions may be purely electronic rather than electro-mechanical. Many combine electronic and electro-mechanical components, as do many of the devices in this book:

1. Opening the firing circuit and short-circuiting the pyrotechnic side is a redundant means of safing.

2. Establishing a circuit to the fuse and removing a short circuit is a means of arming.

3. A plunger or twist-knob generator, at the end of a long wire, provides a firing pulse of electricity for fixed charges only when operated by a human. Non-use of a battery eliminates the possibility of an electrical accident.

4. Powering the arming and fusing system of a bomb by a small propeller-driven generator provides safing. Similarly, powering a fuse by a battery whose electrolyte is separately stored and released at a programmed time and condition is a means of safing.

5. An integrating accelerometer can be used as a stage in arming to allow full arming to take place only after discharge from a cannon or after partial burning of a rocket.

6. Electrical timers can be incorporated into missiles and projectiles for arming or firing or both.

7. An impact fuse can be a piezoelectric crystal or an accelerometer which generate an electrical impulse.

8. A proximity fuse is RF reflection or other field effects in a radio transmitter receiver in the munition.

9. A magnetometer fuse uses a change in magnetic field.

10. A counter in a magnetic mine arms it after several fuse signals, to hamper minesweeping.

11. In naval mines, fuse transducers sense impact, magnetic fields, sound, and pressure waves.

12. Sensing ambient air or water pressure is an arming technique.

13. Sensing spin and the end of spin with a radial accelerometer are arming and fusing techniques.

14. Remote control arming or firing can be done via wire, radio, or sonar.

15. Redundancy is used for reliability.

Enormous amounts of engineering have been expended in this field, but much is kept classified.

15.2 GUIDANCE AND STABILIZATION

Inertial guidance of airplanes, rockets, submarines, ships, and torpedoes uses the EM technology of Chapters 7 and 8. EM gyroscopes sense orientation, and the doubly integrated outputs of EM accelerometers measure distance. Stabilizing the orientation of missiles, airplanes, torpedoes, and tank cannon, without measuring travel distance, requires only gyroscopes [A18, A23, A26, A27, A28, A29, A30].

There are also non-inertial means of guidance using earth's magnetic field sensing, radio (including the satellite Global Positioning System), radar, IR sensing, remote control with TV feedback, laser target designation, altitude measurement, and dead reckoning. The phrases "smart bomb" and "homing guidance" apply.

15.3 SERVOMECHANISMS

Much of what used to depend on human skill and muscle in military operations is now done with instruments and servos. These include aiming, navigating and guiding, and stabilizing vehicles.

Training simulators are not weapons but are major military equipments.

All of the above contain small and large computers, as does logistics management, and we have seen the EM components of computers.

15.4 MAGNETIC RAIL GUNS

Electric cannons used to be the stuff of science fiction, but there is now a real one at the R&D level, described in [Section 6.17].

15.5 EFFECT ON DESIGN

Design of EMDs and everything else for the rigors of military service has had the fallout of new and more reliable civilian products. Many products, techniques, and materials developed on military budgets are directly applicable to civilian use, integrated circuits for example.

CHAPTER **16**

Medical Devices

16.1 HISTORY

In the mid 1950s, the only electrical technology in wide use in medicine was X-ray, diathermy, cardiography, and some laboratory instruments. The electro-cardiographs were of such low bandwidth that later, for a time, some were equipped with low-pass filters so that older doctors could understand the waveforms.

In 1958, a surgeon asked his cousin, an Assistant Chief Engineer at Convair, if he could provide some minor modifications on an experimental heart-lung machine. The final result was a new machine with automatic control of blood oxygenation by a servo varying the oxygen/nitrogen ratio in the "lung," automatic control of pH by a servo varying the CO_2 in the lung, blood-flow metering, and a number of other features.

A hundred successful human operations followed innumerable R&D operations on dogs while the surgeons developed the surgery and the author developed the machine. You can imagine the anecdotes [A3, pp. 14-15; A12]. The machine was finally retired when new commercial machines with similar, and more, features appeared on the market.

Now you can get a Ph.D. in bio-medical engineering and many do. And the FDA would never countenance such informal R&D!

Today, EMDs are used both in diagnosis and in therapy. In popular usage, they are included in "medical electronics."

The devices listed below are EMDs; other uses of electricity and other technologies are not included.

The general requirements for medical EMDs are severe, and many are unique. First, of course, is the ancient medical rule, "Do no harm." Harm can take many forms, including:

1. Electric shock
2. Mechanical injury
3. Chemical or biological injury:
   - Allergy
   - Hemolysis (damage to red blood cells, particularly with large-scale exposure as in artificial organs)
   - Toxicity
   - Carcinogenesis (cancer encouragement)
   - Pain
   - Failure during use
   - New bad effects (induced currents? Other?)
4. Any device which can be on a path that might introduce germs into the body must withstand sterilization. The arrival of AIDS, which can enter the body through an otherwise trivial wound, has magnified the problem.

The Food and Drug Administration (FDA) must approve medical devices and materials, and approval is not easily achieved. Furthermore, the FDA closely monitors manufacturing of devices and materials; a major pacemaker company was shut down for a time because of inadequate quality control procedures. That heart-lung machine, as we tested it and made it, would never get off the ground today.

A second impediment to hasty introduction of imperfect devices and materials is the product liability lawsuit; a class-action lawsuit can cost billions of dollars.

The asbestos case, while not dealing with a material intended for medical use, illustrates a "new, bad effect." Who would have thought that microscopic fibers of a chemically inert material might cause lung cancer? Yet the product liability costs ran into billions of dollars. The current anxiety about possible cancer causing by induction from power transformers is another example of what might become a "new, bad effect."

Yet, so much good may be done by EMDs to come, that we certainly do not want to discourage their development.

16.2 DIAGNOSTIC EMDs

The following are examples of EMDs used in medical diagnosis:

1. Blood pressure measurement, external and internal. Blood pressure cuffs are automatically inflated and deflated for patient monitoring. Cuff measurement is inaccurate because of the elasticity of tissues, so for greater accuracy and for venous pressure, extremely small-diameter pressure transducers are inserted into blood vessels.

2. Blood-flow measurement. This is done ultrasonically by Doppler effect measurement.

3. Electro-cardiograph. The electrodes, electrode cables, and the strip-chart recorder are EMDs.

4. Electro-encephalograph. This instrument measures voltages generated by the brain and appearing on the skin. Its EM components are the same as item 3.

5. Myograph. This instrument measures the voltages of nerves as the voltages appear on the skin. Its EM components are the same as item 3.

6. Ballistic cardiograph. Accelerometers on the patient sense the momentum reaction of heart action on the patient's body.

7. Ultrasonic scanner. A sonar view into many portions of the body, including a fetus.

8. Patient manipulators for X-ray, CAT scan, MRI scan, and surgery position.

9. Film and source manipulators for X-ray and gamma ray cameras.

10. Blood sample manipulators for blood testing. For certain blood tests done in quantity, samples of blood from different patients follow each other down a plastic tube and are analyzed by optical refraction, reflection, spectrography, and similar dry techniques.

11. Electrical thermometer. Analyzing the temperature vs. time curve permits prediction of the thermometer's steady-state temperature faster than waiting for it. A disposable plastic sheath provides sterility.

12. Insertable electrodes and transducers. Different electrode materials produce different electro-chemical voltages when immersed in different body fluids.

13. Cornea electrode. This small piece of metal is temporarily fastened to the cornea. Voltages appear on it in response to certain light stimuli to the eye.

16.3 THERAPEUTIC EMDs

Ever since Galvani stimulated a frog's leg, people have been applying electricity to the human body in the hope of doing some good, and sometimes they succeed. The following are examples of EMDs used in therapy:

1. Small blood vessel cauterizer used in surgery. A small electric arc from an electrode to the blood vessel provides concentrated heat.

2. Defibrillator. In effect, it delivers a one-shot, external, brutal, pacemaker shock with some chance of restarting a stopped heart or one in which the muscle fibers are not in synchronism *(defibrillation)*.

3. Electro-shock treatment applied to the head is an effective treatment for some kinds of mental illness.

4. Diathermy, producing internal heat by high-frequency current.

5. Prosthetics. Once a wooden leg or Captain Hook's hook, passive prosthetics of metal and plastic have been highly developed. *Active prosthetics* respond to nerve signals sensed through the skin; they command prosthetic motions via batteries, amplifiers, and actuators, including small motors. Motion from healthy muscles located elsewhere in the body can either signal or positively actuate prosthetic motions or external machines.
Electrically powered wheelchairs are a form of prosthetic.

6. Intravenous flow monitor. Figure 7-9 shows an IV drop counter which monitors intravenous flow.

7. Intravenous metering pump. More accurate than a gravity and orifice IV, a pump delivers an accurate and stable flow.

8. Lithofracture. Ultrasonic waves in water can be focused on the gallstone or kidney stone of a patient and break it into small particles which can escape without surgery.

9. Exercise machines. Bicycles, stair climbers, and so forth program various loads and record cycles, time, calories, and anything else which might keep the user interested.

10. Ultrasonic cleaners for instruments.

11. Portable recorders. To record transients, miniature electro-cardiographs and other recording instruments are carried by patients walking around in their normal life.

12. Implanted metering pumps. When drugs are taken by pill or injection, their concentration in the blood cycles from peak to valley, rather like the voltage on a leaky capacitor which is periodically recharged. An implanted storage vessel with means to meter out the contents at a uniform rate eliminates this undesirable cycling. The storage vessel is recharged periodically by hypodermic syringe.

13. Hypothermia blanket. A "blanket" with refrigerated water tubes to lower the temperature of patients in surgery.

14. Blood suction pumps and tubes. These remove blood from surgical cavities.

15. Electric and electro-pneumatic tools for bone surgery. Among these are a numerically controlled milling machine for boring the sockets for artificial hips.

16.4 ARTIFICIAL INTERNAL ORGANS

When an organ fails irreparably, it may be replaceable by a transplant or by a totally artificial organ. Sometimes, the artificial replacement need be only temporary during recovery, or during surgery, or while awaiting a transplant donor. Among the artificial organs so far developed are:

1. Artificial kidney. The "dialysis" machine, invented by Dr. Kolff, was the first "artificial internal organ" and the basis for the society of that name.

2. Artificial heart (experimental). Publicized implants so far used cyclic air pressure for pumping power. However, prolonged penetration of the skin with anything ultimately leads to inflammation and infection. AC electric power and signals can penetrate the skin harmlessly via transformer windings inside and outside the skin. Transformer coupling to monitor and to modify pacemakers is presently common practice. Smaller and smaller electric motors are being developed, the smallest being micromachined silicon wafers.

3. Temporary heart and lung machine. This machine is used during open-heart surgery.

4. Cardiac pacemaker. These embedded devices give artificial signals to the heart muscles. Some have feedback to control the signals. The EM problems are:
- The enclosure. It must be chemically compatible with the body. (Titanium is good.)
- The lead wires which connect to the pacemaker and to the heart. They must be chemically compatible with the body and must have sufficient mechanical fatigue resistance to withstand tens of millions of flexures. The answer to that problem is a small-diameter helical spring of high-strength nickel steel alloy [14.2.1, 14.9]!
- The battery. It must be surgically replaced when it runs down, and it had better not run down unexpectedly. Lithium cells are used.

16.5 DATA HANDLING

Immense amounts of both medical and money data are associated with every patient, whether in hospital or out. It is easy to say the word *computer* but despite all the work and investment so far, nurses still spend much of their time at the desk, making paper instead of nursing patients, and doctors' offices overflow with insurance clerks. At the time of writing this, some medical insurance companies consume up to 25% of insurance premiums in their offices.

CHAPTER 17

Consumer Devices

We are all consumers. The readers of this book have enough knowledge of consumer products so that it is unnecessary to describe the individual EMDs which appear in them. However, it is useful to tabulate the products being, or containing, EMDs to remind us of their enormous numbers and variety.

17.1 AUTOMOBILE

The automobile and the airplane have the biggest collections of EMDs. First the auto:

- Power brakes
- Power steering
- Cruise control
- Starter
- Generator
- Ignition
- Fuel pump
- Air-conditioner compressor clutch
- Engine fan clutch
- Locks and burglar alarm
- Tachometer
- Odometer
- Door switches

- Light switches
- Motor-driven seat adjusters
- Motor-driven seat belts
- Air bag accelerometer
- Starter interlock on gear shift
- Steering interlock with starter key
- Panel instruments
- Turn signal switches and relays
- Dimmer switch
- Door chime relay
- Temperature transducers and indicators for
  - Oil
  - Cabin air (and adjustable thermostat)
  - Engine coolant
- Liquid level transducers and indicators for
  - Coolant
  - Oil
  - Windshield wiper fluid
  - Transmission fluid
- Pressure transducer and indicator for engine oil
- Motor-driven accessories:
  - Windows
  - Mirrors
  - Engine cooler
  - Sunroof or convertible top
  - Radio antenna
  - Trunk lid pull down
  - Cassette or CD player
  - Ventilating, heating, and cooling air fans
  - Windshield wiper motor and switches

17.2 AIRPLANE

The passenger airplane, particularly the multiengine jet, has many more than the passenger car. Combining them into sets, we have:

- Engine controls
- Air foil controls
- Landing gear controls
- Navigation controls
- Air-conditioning controls

- Food and drink cooking and temperature controls
- Fire alarm and extinguisher controls
- Entertainment and public address equipment
- Lighting controls
- Airport metal detector
- Airport luggage scales and conveyors

17.3 HOME

In our homes, we have the following devices containing EMDs for programming, driving, controlling, or protecting, sometimes in duplicate:

- Watt-hour meter
- Circuit breakers and fuses
- Light switches
- Light dimmers
- Wall outlet connectors
- Lamp sockets and switches
- Photoelectric switches
- Time switches
- Burglar alarm switches, autodialer, control switches
- Pneumatically operated safety switches for water jets, operated from the bath
- Temperature transducers, timers, control switches, and sometimes motors in:
  - Hot water heater
  - Space heater
  - Air conditioner
  - Oven, grill, toaster, coffee maker, and other cooking devices
  - Dishwasher
  - Clothes washer
  - Clothes dryer
  - Electric blanket
  - Portable electric heaters
- Microwave oven: switches, door interlock switch, fan
- Electric clocks: motor, setting switches
- Automatic sprinkler: timer and valves
- Motor drives for:
  - Can opener
  - Food mixers
  - Food choppers

- Roasting spit
- Clothes washer
- Clothes dryer
- Dishwasher
- Hair dryer
- Ice crusher
- Garage door opener
- Sump pump (controlled by water-level switch)
- Dental water pick
- Electric toothbrush
- Electric shaver
- Vacuum cleaner
- Floor polisher
- Circulating fan
- Exhaust fan
- Air conditioner
- Refrigerator
- Loudspeakers and earphones
- Power tools
- Vibrator
- Audio and video players and recorders, fixed and portable, for tapes, old discs, CD discs
- Necktie rack

- Transducers, media changers, and servos for audio and video players and recorders

- Telephone

- Facsimile machine (FAX)

- Electric typewriter

- Computer
  - Hard disk drive
  - Floppy disk drives
  - CD-ROM drive
  - Loudspeakers
  - Printer
  - Keyboard
  - Mouse

- Still camera and projector

- Video camera

- Satellite antenna

- Electro-static air filter

- Fire alarm

17.4 OFFICE AND STORE

Factory EMDs are covered separately in Chapter 13. This section deals with equipment consumers encounter in offices and stores.

- Bar code readers (at grocery checkout counters and libraries)
- Photocopiers
- FAX machines
- Credit card readers and bill printers
- Cash registers
- TV camera pan and tilt
- Stolen merchandise exit alarm; book insert magnetizer and demagnetizer in libraries
- Automatic door openers

And they keep coming!

CHAPTER **18**

Passive Electrical Components

In circuit design and analysis, a passive electrical component is merely a parameter (resistance R, self-inductance L, capacitance C, mutual inductance M) in an equation. The rest is mathematics. But actual components are complex EMDs.

18.1 RESISTORS

The heart of a resistor is merely a piece of metal or carbon. But that heart must be supported by insulation, it must have terminals to connect it to its circuit, it may need a package between it and its environment, and it may need a sliding contact for adjustment. It may have taps between its ends. Above all, it must have means to dissipate heat; in fact, its function may be as a heater. It may be a rigid body or it may be flexible to conform to a member it heats (as in an electric blanket).

Most metal resistors use nichrome because of its low temperature coefficient of resistivity and its high-temperature corrosion resistance. Some use stainless steel and some use other alloys. Some tungsten filaments in incandescent lamps are used as resistors, and some of those are designed as radiant heaters.

Resistors in low frequency circuits have few electrical requirements on their shape, but in high-frequency circuits, skin effect confines the current to the region near the surface. In microwave systems, resistors and attenuators must be only a thin film. In stealth military airplanes and submarines, the skin is made with resistance in depth to absorb radar microwaves.

Many resistors are made as coils of resistance wire. Such coils have inductance which may not be permissible. Non-inductive resistors have two concentric coils in series, connected so that their ampere turns are in opposite directions and cancel each other.

Resistors are made in wide spectrums of magnitude and accuracy. Precision resistors are typically non-inductive and are made with careful attention to manufacturing details so that small effects will not change their resistance over time.

Resistors which must dissipate many watts—or kilowatts—operate at high temperature and use ceramic insulation. When it is permissible, their resistive elements have minimum contact area with their supports and maximum contact area with the atmosphere to which the heat must be transferred. Conversely, some exposed resistance heaters, such as for domestic electric stoves and for immersion in liquids, embed the resistive element in ceramic powder and jacket the combination in refractory metal.

Water solutions are sometimes used as resistor elements, particularly when the purpose of the resistor is to boil the water.

18.2 CAPACITORS

A capacitor is a geometrical anomaly: It must pack a maximum of area into a minimum of volume. Most capacitors comprise a four layer sandwich: a ribbon of dielectric with a thin foil of metal on each face topped by a second ribbon of dielectric on one face. The four-layer ribbon is spiral wound into a solid cylinder. The cylinder may be round or flattened. The metal is usually aluminum.

The dielectric may be solid or porous. If porous, the entire winding is immersed in a liquid dielectric with better constants than is available in the solid. (The winding is evacuated before immersion to prevent trapped air bubbles, just as coil windings are evacuated before liquid impregnation.)

It is a problem to make good connections to the metal foils. They are very thin and may be very long, so that their ohmic resistance from inner end to outer end of the spiral may be substantial. In some cases, even their inductance may be undesirably high. The usual solution is to extend one foil to one edge of the dielectric ribbons and the other foil to the other edge. Then a conductive layer is deposited on each end of the cylinder so that the electrical path is axial, is only as long as the cylinder length, and is as wide as the entire spiral. This construction is particularly important if the capacitor is to be used as a source of high-power, short-time, pulse discharges.

Electrolytic capacitors have a large surface area generated by etching miniature hills and valleys on the surface of a relatively small piece of metal. The dielectric is generated chemically as an oxide layer on that metal, only a few molecules thick. The second plate may be similarly treated to make a symmetrical capacitor, and a conducting liquid extends conductivity to the outer surfaces of the two dielectric layers. In effect, there are two capacitors in series.

If the second metal plate has no oxide layer, it is merely an electrode terminal for the liquid and there is only one capacitor rather than two in series. However, this capacitor can be used only with DC of one polarity or the second plate will become oxidized.

Adjustable capacitors of small capacitance for adjusting RF circuits are made of two interleaved stacks of metal plates, with small air gaps, with the area of overlap mechanically variable.

Very small capacitors are made as alternating flat layers of metal and ceramic.

18.3 INDUCTORS

18.3.1 Windings

Inductors vary from a few small turns of fine wire (for a radio receiver) through a few large turns of heavy tubing (for a radio transmitter or a utility system), to a ton of wire on an iron core (for a power transformer). Inductors may be two-terminal self-inductors, three-terminal autotransformers, or transformers with two or more separate windings, each with several taps.

The windings are almost always of copper, although during World War II, when there was a copper shortage, the government borrowed from its silver reserves and some large transformers were made with silver windings.

Transformers are wound for voltages close to a million volts for high-voltage transmission lines. Insulation from winding layer to winding layer must be substantial, and even turn-to-turn insulation at the winding ends must be great, to prevent arcing from transients. In large power transformers, electro-mechanical force on turns may be substantial, particularly during transients and short circuits, so the turns must be strongly supported.

The capacitance from winding layer to winding layer may be a significant circuit parameter. This is particularly true in data transmission circuits where a transformer may be a source of crosstalk and hum. Great ingenuity has been expended in shielding windings from each other.

18.3.2 Cores

Iron cores of several types are used to increase the self- and mutual inductances of windings:

1. *Toroids* (doughnuts) are made either of ferrite or of spiral wound alloy strip, such as permalloy. The winding, either a single inductance or a transformer, is distributed uniformly around the core. There are no air gaps in the magnetic circuit, and all the core material is oriented parallel to the flux. Leakage flux is zero if the windings are uniformly distributed.

Special machines have been developed to wind coils through the holes in the toroids. These machines are cousins of the special machines developed to wind coils in the slots of motor or generator stators and rotors.

2. A special case of toroid inductance was core memory for computers, now made obsolete by semiconductor memory. The toroids were tiny ferrite rings supported in multiple planes of row and column arrays. The windings were long wires threading rows, columns, and diagonals, effectively linking each toroid with three separate, single turns. Memory was retained by the direction of residual magnetism in each core.

3. Another special case of toroid is the smoothly adjustable autotransformer. A single winding extends about a toroid. The winding's outer surfaces on one end of the toroid has its insulation removed, and a carbon brush is pressed against it. The angular position of the brush is adjustable. The voltage from an end terminal to the brush is an adjustable fraction of the voltage between the end terminals. The core is a stack of stamped toroidal laminations of transformer steel. The device is analogous to a resistance potentiometer except that its only losses are those of a transformer plus a small resistive loss in the brush. The brush is carbon, like a commutator brush, to limit the losses in the turn it short-circuits.

4. Flat lamination stacks. Magnetic laminations shaped like the letters I, U, and E are inserted through the coils and stacked into solid cores. Alternate layers are reversed, so the air gaps of each layer are bridged by continuous metal in the adjacent layers.

Some inductors have short air gaps deliberately left in the lamination stacks, and some have adjustable air gaps to produce adjustable inductors.

5. Some power transformers use a continuous strip of silicon steel spiral wound through the coils. This construction eliminates all air gaps and aligns all the steel's oriented grains in the direction of the flux.

6. Variable self-inductance and mutual inductance is achieved by a movable ferrite slug on the axis of otherwise air-core coil or coils. The LVDT transducer described in Chapter 7 is made in this way.

7. Some three-phase power transformer cores have only three parallel legs bridged together at each end. Each leg has the other two legs as a magnetic return path. If you write the equations for the fluxes, it works out as elegantly as the equations showing why three-phase electric power needs only three conductors.

8. The current scare about cancers being induced by power transformer leakage flux will probably result in modified configurations which will be somewhat less cost-efficient but which will produce less leakage flux.

18.4 BATTERIES

Leaving the semantics aside of whether or not batteries are "passive components," they certainly are EMDs and this seems a good place to mention them. Both primary batteries and storage batteries are complex electrochemical-mechanical devices. Fuel cells are a form of primary battery in which the consumable chemicals are replenished just as, in storage batteries, the electric charge is replenished [1.10, 15.1–4, 16.4]. The subject is a large one, and this book contains only these brief mentions.

18.5 ACTIVE COMPONENTS

Semiconductor design is outside the scope of this book, but semiconductor manufacturing is surveyed in Sections 2.11.1 and 2.11.2. Vacuum tubes are also omitted from this book, although a vacuum tube is a complex mechanical structure enclosed in a gas-tight envelope. Both classes of components are presented in great detail in other publications.

PART III

Understanding Design

The next chapters are essays on aspects of design, primarily relevant to designing EMDs, but covering principles which apply to all designing and to the careers of design engineers. These essays are not a textbook of design, but they contain ideas useful to designers and to the understanding of EMDs by those who do not design them.

As you read, please note the designer's constant need for *judgment* as well as for technical knowledge and mathematics. To some degree judgment is a talent, and to some degree it is learned from study and from experience. But remember engineer Alpha who has 20 years of experience and engineer Omega who has been doing the same thing for 20 years.

CHAPTER 19

The Science and the Art

The *science* of engineering design is physics, chemistry, and mathematical analysis of devices and systems. Materials may be considered a branch of chemistry, and measurement may be considered branches of all. Engineers use the science to *predict performance* and to *size parameters*.

The *art* of engineering design is the knowledge of everything else which can be useful in design and the skill to use that knowledge. The following chapters are devoted to phases of the art.

The science is quantitative; the art is qualitative. But the science also crosses over the boundary by teaching *insight,* that is, *intuitive understanding,* of the behavior of matter and energy. This insight enables design engineers to imagine and understand the behavior of existing and proposed devices and systems to a degree that those untutored in the science cannot match.

19.1 THE SCIENCE OF DESIGN

Most of our academic engineering study is of the science, and it should be. Mathematical physics and engineering require disciplined study and must be learned in sequence. One cannot pick up a little calculus here and a little vector analysis there and accumulate either understanding or utility. And the one sure thing about using cookbook formulas is that they will be used incorrectly. But our study of the art *can* be piecemeal and can go on throughout life.

It is easier to learn the science of an engineering phenomenon if a non-mathematical description of the corresponding art is given as an introduction. Most books on engineering design are books of parameter relationships and calculations. Stress, voltage, strength, impedance, deflection, oscillation frequency and amplitude, temperature, and other parameters are their subjects for mathematical analysis. *This* book assumes that engineers learned these relationships and calculations in their academic courses and that non-degreed designers take difficult calculation problems to engineers.

Computers are mathematical instruments of enormous value in the *science* of design. Furthermore, one of the graphics capabilities of computers is drafting; computer-aided drafting is an extremely valuable tool in the *art* of design. (The ambiguous acronym CAD can mean either Computer-Aided Drafting or Computer-Aided Design; it depends on the capability of a particular program as to how much science it can provide. The expression Computer-Aided Engineering, CAE, clearly implies both.) See Reference [A36] for a comprehensive study and reference [A37] for a balanced discussion on the benefits and malefits of computer-aided drafting.

19.2 THE ART OF DESIGN

Your study of the art of design began in your childhood with toys and tools and the devices of life. It should never stop until senility turns off your mind. This is not poetic exaggeration; any bit of knowledge may be useful when least expected.

Permit an anecdote to illustrate: Rabinow Engineering Co. developed the first automatic mail sorter for the U.S. Post Office. The author's manager told him to design a code carrier to escort each moving letter, and a code detector for each stationary sorting bin. Both carrier and detector had to be all-mechanical because the Post Office had no electronic technicians at the time. Since there would be approximately ten thousand code-bit sensings per second, the sensing action was to be impact free to minimize noise and wear.

After a long and heated argument, the author said "It can't be done and I can *prove* it mathematically!" The manager said, "OK, it can't be done, but if you *could* do it, *how* would you do it?" Since they were both purple from shouting at each other by this time (you know, of course, that design engineering is a dispassionate and totally rational process) the question seemed perfectly reasonable at the moment. The author looked up at the ceiling and saw a variation of *an image in a Bob Hope comedy song in a recent movie.* He sketched it, whereupon the manager really got mad and demanded, "If it was so easy, why did you give me such a hard time?" (There are millions of these elements now in service.) [Section 8.5.5.2 and Fig. 8-8.] Photographs are on p. 21 of [A3].

The moral:

1. *All* knowledge is grist for your mill.
2. Be very careful about what you say can't be done so you won't have to eat your words.

Solving problems by new ideas and by choosing among many kinds of apples-or-oranges options cannot be quantified; you are left with only your qualitative knowledge, ingenuity, and judgment to work with. *Judgment* is a word which will arise frequently in the remainder of this book. It is in the realm of *ideas* rather than *numbers* that the art lies.

There are many, many aspects of design which are not subject to calculation at all. Consider *robustness, aesthetics and customers' and managers' tastes and prejudices,* which are only three of many. Some of these non-mathematical components of design are covered in the following chapters.

A vital portion of the art of design engineering is a knowledge of manufacturing technology and economics. One cannot design a part or assembly without knowing how it can be made (there are jokes about parts which *cannot* be made) and at least the relative costs of making it in different ways and in different quantities.

The following chapters discuss aspects of design [A3, A4, A6, A40, A41, A42, A43, A44].

CHAPTER 20

How One Designs

20.1 CALCULATING A DESIGN

One cannot "calculate a design" with or without a computer. One can only calculate the *magnitudes of the parameters* of a *qualitative* design to convert it into a *quantitative* design. The numbers may then suggest changes in the qualitative design, and one then iterates.

Inventive design always starts as qualitative ideas, even when those ideas are based on educated insight into mathematical relationships. (However, in the real world, much design work is scaling up or down an existing design or otherwise adapting it to new specifications, and much of that work is calculation.)

One *can* initiate some designs with a quantitative study to extend performance specifications into device parameters, for example, power, enclosed volume, speed. Such a study can sometimes be expressed as families of curves which guide the designer to start with some parameters at almost final size.

20.2 THINKING A DESIGN

There is no systematic thinking procedure in inventive design, although some theoreticians keep trying to organize it into a systematic procedure or even a computer program [A3, Chapter 28]. New thoughts feed back to change preceding thoughts. You must think of all considerations at the same time; *thinking must be parallel rather than serial.* You consult others *(Concurrent Engineering* and the like) and they add to the conflicting ideas. You delegate part of the work to subordinates, and when you see their work, you decide to make changes, and they say "Why didn't you tell me what you wanted in the first place?" (The answer is that if you were infinitely smart you would have. Drafters have an occupational disease called "indelible pencil.")

You design from the outside in and from the inside out. You are in constant mental turmoil. If you have coworkers that turmoil appears among you. With mutual respect it is constructive, as in brainstorming sessions, but with tender egos, conflicting political motives, and disrespect, it generates quarrels. There is continuing change until the freeze bell rings. If you worry that you have been doing something wrong because your thoughts were not proceeding in an orderly fashion, accept consolation; that is the way a mere mortal designer works.

This is not to say that designing is totally chaotic. Where alternatives can be defined and quantified, quantitatively optimum choices and combinations can be computed. References [A2, A50] and their references present such calculations. Typically, a decision matrix is formed with a point system and loading factors. The matrix is solved by a computer (or just on a blackboard) to find an optimum. One must beware of the temptation to weight the loading factors in favor of a desired conclusion.

20.3 RULES FOR DESIGN

One can recite a variety of *Rules For Design,* but most are obvious, such as the "rule" that one should study the specification before starting to design. This book will discuss principles rather than "rules." The only serious "rule" is, "It Depends."

Here are a few principles which may help you:

1. Some of your design ideas and decisions will be based on tests and experiments. We have all had college training in scientific method, but there is a tendency to rush things and conduct tests and experiments in a hasty and careless fashion. The results can be costly and misleading. Please take the time to measure, to observe, to hypothesize. Design and perform experiments to test your hypotheses and to record data. Try to change one thing at a time.

Much of our debugging work is debugging the bugs we put in to overcome or test the original bugs.

When a device acts improperly—jams, for example—there is an overwhelming compulsion to reach out and clear the trouble. DON'T DO IT! If you do, the fault *will* occur again, perhaps very much later, in actual service. *Look* at it. *Measure* it. Devise an experiment to *analyze* it. Do not proceed until you *understand the fault* and correct its cause.

2. In choosing design alternatives, predictability is a value. A reliably predictable design may be better than a more elegant design whose performance you cannot reliably predict. The choice may depend on the available testing budget.

3. Get quantitative specifications if you can. Qualitative requirements cause fights when acceptance time comes around (for example, "Good commercial practice").

But if you design well, you may see the results in actual working hardware, and it all becomes worthwhile!

20.4 THE VOCABULARY OF DESIGN

Just as a writer has a vocabulary of words, a designer has a vocabulary of materials, parts, part features, and assemblies. The writer selects and combines the items in his or her vocabulary into the writer's text; the designer selects and combines the items in his vocabulary into the designer's design. The richer each vocabulary, the richer each product. *Expanding your design vocabulary is a lifetime program.*

The design engineer's vocabulary includes:

- Passive electrical components (R, L, C, diodes, and so forth)
- Active electronic components (transistors, ICs, and so forth)
- Mechanical components (fasteners, gears, and so forth)
- Electro-mechanical components (relays, connectors, and so forth)
- Fluid power components (valves, pumps, and so forth)
- Materials (metals, plastics, and so forth)
- Manufacturing tools and techniques (machine tools, measuring instruments, and so forth)
- Part features (threads, bosses, and so forth)

Writers use standard ("boilerplate") paragraphs preassembled of many words, sometimes filling in blanks for the immediate use. Designers use standard assemblies of many components, sometimes filling in parameter values. Examples are:

- Amplifiers
- Motors
- Hydraulic power supplies
- Electrical motor controls
- Electrical program controls
- Pneumatic controllers

Some writers have rather standard stories into which they fit new characters and locales. Some designers have standard products into which they fit new components, sizes, and design details. Examples are:

- Radios
- Engines
- Cars
- Airplanes
- Ships
- Computers

(This would be a good point to lighten up and consider the difference between a *component* and a *system*. Everyone buys components and sells systems. "System" carries more prestige than "component." For example, that simplest of components, a cheap carbon resistor, is actually a rather complex system comprising resistance element, leads, connections, case, and marking. On the other hand, to a fleet admiral, an aircraft carrier is merely a component.)

An engineering dictionary is an encyclopedia. Engineering encyclopedias are listed in [Group C]. Handbooks [Group B] serve as small encyclopedias in addition to having numerical data and design formulas. Unfortunately, a dictionary does not help unless you already know of the word, either because you have just seen someone else's use of it or you already know something about it.

The best detailed encyclopedia for manufactured items which interest you is their manufacturers' catalogs. Many are instruction manuals explaining the products and teaching how to specify and use them. [A35] is an example.

The index for your catalog encyclopedia is *Thomas Register* [E1]. It is the "yellow pages" for all products made in America. It also has a trade name list that lets you trace a product to its manufacturer and information about the manufacturers of the products it lists. *Thomas* fills many volumes. Other directories are named in [Group E].

The catalogs in [Group F] are particularly useful for small-lot manufacture since they contain uncommon components expensive to manufacture in small lots but inexpensive to buy from the vendor.

It is useful to compile your own catalog collection and to file it in library boxes by subject rather than by vendor's name. You can page through for related ideas without jumping from vendor name to vendor name, directed by an index.

A good way to expand your design vocabulary, and to exercise your judgment, is to observe and criticize the designs you encounter in daily life. Some you will admire and plan to imitate. Some you will find fault with and will design improvements of, at least in your head. Pay particular attention to foreign designs; you will find that there are tacit assumptions in American designs which are not made abroad, and you will learn new ideas from the foreign approaches. Reference [A5] is a critique of common designs; it is also great fun to read.

There are many sources for the continuing expansion of your design vocabulary. One is the design magazines [Group D]. These journals are free because they are paid for by the advertisers. Unlike most consumer advertising, many of the ads are highly informative and invite requests for detailed literature via their "bingo cards."

One of the best ways, and certainly the most interesting way, to improve your design vocabulary and your design judgment is to attend trade shows. There, you will examine actual hardware. You can question its sales engineers and sometimes its design engineers and company officers. You can pick up or ask for catalogs. You can look for a better job. If you give a technical paper, your employer may subsidize the visit for the publicity he gets. If your employer is an exhibitor, and if you can get to do booth duty, you will find time to visit other booths and to do your own market survey on your company's products.

The bad news, of course, is the cost in time and money, your need for employer's permission, time, and money, to say nothing of crowds, tired feet, and bad food.

20.5 PLASTICITY

Designers have a vital freedom where the writer simile ends: In three dimensions we stretch, shrink, displace, deform, configure, proportion, and otherwise create structures and relationships with no direct analogy in prose. Sculptors and painters have similar freedoms; the word *plastic* is used by art critics to describe such artistic design. In this sense, the work of an inventive designer is plastic.

20.6 CONSTRAINTS

Prose writers are constrained by rules of syntax which derive from custom; designers are constrained by the laws of nature, available technology, receptivity of the marketplace, contracts, specifications, and laws.

Writers, poets, and plastic artists are free of laws of nature but are constrained by the tastes of the public they hope to please and sell to. Designers are constrained both by the laws of nature and by the tastes of their managers and customers.

Chapter 30 is devoted to your constraints as a designer.

20.7 THE DRAWING BOARD AS LABORATORY

After your rough sketches, your first layout drawing is your first trial design of an EMD; it is an experiment to help you think and visualize to scale. When you study the drawing, you decide on changes for innumerable reasons and put them into your second layout. In fact, you start the changing process almost as soon as you start the layout. Even after you are satisfied, you must show your layout to your managers and perhaps to your customer before you are authorized to proceed with detail design.

The drawing board and the CAD terminal are design laboratories first; then they are used to make manufacturing drawings.

Regardless of your rank, you should make the first design layouts with your own hands and not instruct a subordinate with freehand sketches, arm waving, and words. Any intermediary will limit your freedom to think and change. It is not uncommon for a company president to have a board or CAD terminal in the president's office.

CHAPTER 21

Minimum Constraint Design

21.1 IMPORTANCE

- Minimum constraint design (MCD) permits the design of mechanisms and structures with zero looseness, zero binding, and low cost.

- MCD is insensitive to loose manufacturing tolerances and to dimensional changes due to foundation settlement, temperature changes, and the like.

- MCD permits the design of stationary structures with no forces among ground, assembly, and subassemblies other than those necessary to provide support.

These are big claims.

Their truth was demonstrated in the design, manufacture, and installation of large Cartesian robots by MOBOT Corporation between 1980 and 1985. The last machine made during that period handled 300-pound lathe fixtures holding jet engine parts, carried them along a 400-foot row of NC lathes, and loaded and unloaded them into spring chucks with 0.015 inch clearance when open. To make life more interesting, this robot was mounted on building columns which deflected under wind load, and the lathes were on a foundation that was slowly settling.

At Numerical Control Corporation, engineers developed and used the technology in both small precision devices and in large machines.

21.2 PUBLICATIONS

Our original source of the concepts in MCD was the book, unfortunately now out of print,

T. N. Whitehead, *The Design and Use of Instruments and Accurate Mechanisms* (New York: Dover, 1954).

MCD theory was first published by McGraw-Hill in the book *Successful Engineering,* republished by IEEE as *Real-World Engineering* in 1991 [A3]. It was then much further developed in *Designing Cost-Efficient Mechanisms,* published by McGraw-Hill and republished in 1993 by the Society of Automotive Engineers [A8].

21.3 THE TECHNOLOGY

An introduction to the technology is best given by its condensed description in *Real-World Engineering,* Chapter 31, "Minimum Constraint Design (MCD)."

21.3.1 What It Does

This chapter deals with design principles for achieving, simply and economically, mechanisms with zero looseness, zero binding, and zero stress due to assembly. The same principles produce static structures which can be assembled with zero looseness and zero interference, due to tolerances, during assembly. The principles are old, but I have met few mechanical engineers who are acquainted with them. I have used these principles for years and trained my staffs to use them, with great benefit to our companies and clients.

Whitehead, who was my source of education in the subject, uses the expression *kinematic design* for these principles. This expression suggests the design of moving linkages, which it does not mean here, so I use the expression *minimum constraint design,* which I think is more descriptive.

21.3.2 The Basic Theory

An unconstrained rigid body has six degrees of freedom which I will call X, Y, and Z linear freedoms and Roll, Pitch, and Yaw rotary freedoms. If it is constrained at 1 point, it retains 5 degrees of freedom; if it is constrained at 2 points, it retains 4 degrees of freedom; and so on until if it is constrained at 6 points, it has zero freedom. If an attempt is made to constrain it at more than 6 points the body will be forced to deform, that is, there will be binding or interference or internal stress from the deformation needed to conform to the overconstraint. This sounds pretty abstract, so let's take an example.

21.3.3 An Example

In Fig. 21-1 rigid body A is mounted on rigid body B on 3 legs, C, D, and E. Each leg has a hemispherical tip. Assume that all parts and shapes are made with loose manufacturing tolerances.

Leg C seats in conical hole 7. Since the leg end and the hole both have finite shape errors, they will touch at 3 points: 1, 2, and 3. If you imagine holding A in your hand, you will see that A now is constrained in X, Y, and Z but it remains free to move in Roll, Pitch, and Yaw. Three points of constraint have removed 3 degrees of freedom.

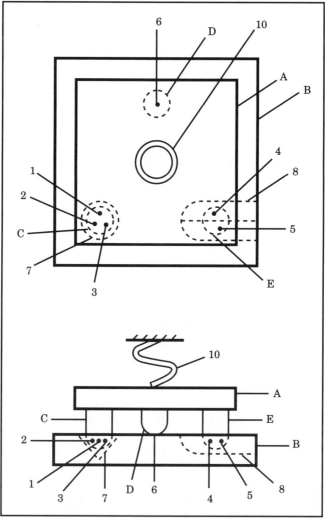

Figure 21-1   Two-Body MCD Assembly

Now seat leg E in V-shaped groove 8. It will touch at points 4 and 5. Body A now has only one free rotation left, about the axis joining the hemispheres at the ends of legs C and E. Five points of constraint, 5 degrees of freedom constrained, 1 degree of freedom unconstrained.

Now rotate A about that C–E axis until leg D touches the flat surface of B. Six points of constraint, zero degrees of freedom.

Add spring 10 to hold the two parts together. The spring provides a *force* but does not add a *constraint* because the spring end is flexible in all 6 axes.

I have built working devices with exactly this structure, and the effect is weird. The parts go together and come apart with no effort whatever, yet when they are together, they have zero looseness and feel as if they were welded.

Compare this with a conventional doweled assembly.

21.3.4 Heavy Loads

Lightly loaded structures can be made in exactly this way, but heavy loads will Brinell the points of contact. This may not be bad—the cold worked contact areas will still be nearly perfect in effect and yet have finite contact areas. Even lightly loaded "point" contacts elastically form contact areas, as in a ball bearing or electrical relay contact.

For heavily loaded structures you compromise with the theoretical purity of your point contacts in a variety of ways. The cone seat 7 can be a spherical seat of slightly larger radius than the C leg end, so the contact stresses under high load are tolerable. The tip of leg E can be wedge-shaped and free to rotate about the E axis to give a comparable effect. The tip of leg D can sit on an intermediate block with a spherical hollow on top and a flat below with the same effect.

You can devise many other forms of replacing the theoretical 6 points with 6 areas and still have the same elegance of zero looseness, zero binding, and loose tolerances.

If you are designing a mechanism with moving parts, you do exactly the same thing but you use 5 or fewer constraint points, depending on the number of degrees of freedom you want for each part.

21.3.5 Using Commercial Components

A very valuable exception to pure minimum constraint design is to make the interfaces between bodies commercial components made with close-tolerance, multiple-internal constraints. Among these are:

- Rolling contact bearings
- Spherical rod ends

- Precision ways with matching linear self-aligning bearings
- Tooling balls and separate balls
- Spherical washer pairs
- Tooling pressure pads with internal spherical joints
- Universal joints
- Flexible couplings

The only field of design which I know that uses minimum constraint practice regularly is machine-shop fixture design. The fixture design (or tooling design) component companies make many products which can be used in a variety of structures and mechanisms to provide small area constraint "points."

The author had an amusing consulting experience using this art. I visited an electronics company which had mounted an instrument servomechanism on the same board as its control circuitry. The rotating parts were very tight (that is, binding), and their technicians could not make them turn freely. Cables were held down with many cable clamps and, to match, shafts were held down with many bearings. I told them to do nothing but remove certain bearings. Instant fix. Client awe.

In MOBOT Corporation, we made machines with 150 feet of horizontal travel and 20 feet of vertical travel using only a few simple machine tools and no machining of large areas. We shipped without fully assembling them in our own plant because our shop was too short. They went together in the field without trouble and with zero looseness and zero binding.

The author has found that the use of these principles is a skill which grows with practice. He strongly recommends that you do practice it. You will achieve both functional and economic success.

CHAPTER 22

Design for Manufacturing

22.1 WHY

Everything you design must be manufactured; other things being equal, the easier and cheaper it is to make your design, the more likely it will be used and the higher will be your standing and your satisfaction.

You can call for an unprecedented design if you can devise a method to make it. Development of product and process together can produce advanced and valuable products.

The mechanics in your shop can be a valuable source of knowledge. Many an old-timer has seen much that is not in a textbook and may have developed an excellent judgment. He/she may not really understand what he/she tells you because he/she does not have your education in theory, but you can interpret what he/she says. He/she may make specific suggestions to improve your design.

(He/she may also make changes without your permission, to "make it work," thus spoiling your documentation and maybe improving—or maybe spoiling—your product. You must watch the creative mechanics like a hawk, and be tactful about it, too!)

22.2 QUANTITY

Some processes require large tooling cost but have low part cost: injection molding, die casting, and large-part stamping are examples. Since total part cost includes tooling amortization, such processes are permissible only for large-quantity production. But these processes may also produce parts available in no other way. (A serious problem in using such processes is the cost in money and time of making models.)

Other processes, such as drilling holes, require little or no tooling but may have large part manufacturing cost.

A real-world problem is the need to produce the same part in quantities of one for development, small lots for testing and early marketing, and large lots for final production. There are techniques, such as short-run stamping, for short runs and it is well worth your while to learn them. New prototyping and small-lot manufacturing processes are always being developed. 3D photolithography is one of the more recent. Their use may spell the difference between success and failure of a design approach [A47].

22.3 PREFERRED PROCESSES

You should learn your company's manufacturing preferences or you will be pressured to make changes after you disclose your design. Does your company prefer to manufacture in-house or does it like subcontracting? What are the capacities and accuracies of its equipment? What is the spectrum of skills of its people?

The answers may not be easy to determine; different departments and different people may have different preferences. When in doubt, ask your own manager what to do. And hope he knows.

If subcontracting to specialized vendors is permitted, you have an enormous wealth of manufacturing technology available to you, together with free advice from the vendors on how best to use it. The price to you is learning what is out there. You should read the ads in design magazines as a regular practice [20.4].

22.4 MATERIALS

There is no field of engineering in which as many kinds of material are called for as in EM design. This section describes a source reference for EM designers.

The Annual Materials Selector Issue of *Machine Design* magazine [B16] contains a set of tables which compare the properties of different materials. These "comparison of materials" tables help select the best material for each part. Each table is ordered by the magnitude of its parameter, not by alphabetical order. To emphasize the point and to encourage you to get the book, here is the list of its comparison tables:

- Coefficient of thermal expansion
- Density

- Dielectric constant, Non-metallics
- Dielectric strength, Non-metallics
- Electrical resistivity
- Elongation of metals
- Hardness of ceramics
- Hardness of metals
- Hardness of plastics
- Hardness of rubber and elastomers
- Heat deflection temperature of plastics
- Maximum service temperature of non-metallics
- Modulus of elasticity in tension
- Specific heat
- Specific stiffness
- Specific strength
- Tensile yield strength
- Thermal conductivity
- Ultimate tensile strength

It is sometimes useful to construct a short table of candidate materials by defining a figure of merit for your particular problem and determining the figure for the different candidates from these tables.

There are many other publications on materials; some of them are listed in the References [A7, B8, B10, and most of the handbooks and encyclopedias in Groups B and C].

22.5 PROCESS CATEGORIES

The field of EMDs is so diversified that almost all manufacturing processes are used to make them. The following is a very brief review of categories of processes used in the manufacture of EMDs. This is only a category list covering a subject requiring entire books for complete coverage. See references [A2, A4, A17, A45, A46, A47, B5, B18].

1. Convert amorphous material to parts (Metals, plastics, elastomers, ceramics and glass)

2. Convert mill products to parts (Flats, rounds, tubes, rolled shapes, extruded shapes, drawn shapes). Those processes which apply to mill products also apply to castings, moldings, and extrusions as second operations.

3. Bulk processes, for example, deburring, heat treatment.

4. Surface finish, for example, electro-plating, painting.

5. Inspection. (Inspection is a manufacturing process which you should include in your design plan. Inspection should be easy to do and should be possible at part, subassembly, and final assembly stages so that faults can be detected when it is least costly to correct them. It is useful to design in mechanical and electrical features as fixture and instrument interfaces to make inspection easier.)

6. Permanent assembly, for example, rivetting, welding.

7. Non-permanent assembly, for example, threaded fasteners.

8. Selective assembly (by grading parts).

9. Subassembly. (Subassemblies are useful even for single quantities of large products, such as ships. Accuracy is required so details fit into subassemblies and subs fit into supers.)

22.6 DESIGN FOR AUTOMATION

See [B5] for detailed descriptions and different varieties of these processes and for design rules for each.

1. Analog template and cam control:
  - Machine tools
  - Contour burners (flame, arc, laser cutting)

2. Numerical control (NC, CNC):
  - Machine tools
    - Tool changers
  - Contour burners

3. Robots
  - Machine loading/unloading
  - Fabricating
    - Spot welding
    - Arc welding
    - Part manipulation
    - Spray painting
    - Adhesive and sealant deposition

4. Automatic material handling:
  - Conveyors
  - Automatic guided vehicles (AGV)
  - Automatic storage and retrieval systems
  - Machine tool cutter magazines

5. Transfer machines:
- Fabrication
- Assembly

6. Automatic testing machines

22.7 DESIGN FOR AUTOMATIC ASSEMBLY

1. Replace fasteners with:
- snap assembly
- deformation of parts
- press fit
- spin
- spot or arc weld
- ultrasound bond
- toy tabs
- adhesives

2. Design parts for easy and reliable feeding by part feeders.

3. Design for assembly by simple machine motions instead of by human dexterity.

22.8 SHORT-RUN MANUFACTURING

The product may be required in quantities of one article and up, either because that is the actual need, or because more test models are needed, or because they will be used for test marketing, or for other reasons.

Short runs impose severe constraints on the tooling investment permitted, and they in turn impose severe constraints on the part designs permitted.

Because there is such a need for short runs, many ingenious techniques have been developed to manufacture them.

The art of manufacturing is part of the art of design engineering.

CHAPTER 23

User-Friendly Design

The most felicitous phrase in engineering's vocabulary is "user friendly." It implies all that is good in form and function. Let us consider both.

23.1 FUNCTION

The simplest rule to follow is to imagine yourself using your own product. Since each of us is prejudiced in favor of our own designs, the next thing to do is to observe someone else using a development model. Ask yourself these questions:

1. Is it easy to make errors due to misleading appearance or touch?

2. Is it fatiguing to use, either in general or in some particular way? A common effect of certain button-punching machines is carpal tunnel syndrome, which has led to many sore hands and wrists and many product liability lawsuits.

3. Visibility during installation, use, and maintenance: Is there glare? Obscuring shadows? Poor accessibility?

4. Is it resistant to damage from negligent or malicious use?

23.2 SPECIFIC COMPONENTS

1. Knobs and handles should be comfortable to the touch, particularly when substantial force must be applied. They should be recognizable to both the eye and to the touch to avoid using the wrong one. Shape, color, words or symbols, and location provide redundant identification to the eye. Shape, surface texture, heat conductivity (metal feels cool, and so forth), and location provide redundant identification to the touch. The shape and labels of the knob or handle should suggest its motion and function: rotate, pull, push, hot, cold, and so forth.

2. Avoid sharp corners and edges, which can hurt, and hook shapes, which can snag clothing.

3. Provide body comfort (cushions, seat height, height from the floor for maintenance access, illumination, ventilation, and so forth).

4. Prevent the turning of a terminal screw from bending its wire or splaying the strands. Use any of the many techniques in commercial practice.

5. Follow conventions. Such matters as door-swing direction, valve-turning direction, and sign location are the few places where creativity is a bad thing.

6. Make it a habit to criticize the human-made world around you for user friendliness. You will automatically become better at it in your own work.

The book *The Design of Everyday Things* by Donald Norman [A5] is both entertaining reading and a serious discussion of these matters. Reference [B21] and newer books in the field of *Human Factors* or *Human Engineering* expand the ideas merely surveyed in this chapter into a professional study.

23.3 FORM

An ugly product that works will not sell as many as, or be treated as carefully as, a good-looking product that also works. There is a profession called *industrial design* which modifies designs which only work into designs which also look good. Industrial designers also suggest details of user friendliness such as those above and add an element of human taste in form and beauty.

Big companies may have industrial designers on staff; small companies may call them in as consultants. They are yet another source for concurrent engineering, but the sooner they express their opinions, the less painful and costly it is to benefit from them.

23.3.1 What to do Until The Industrial Designer Comes

The following is a list of suggestions for do-it-yourself industrial design. A professional will sneer at your results as amateurish but the changes he/she makes will be much fewer and less exasperating than if you do nothing. Furthermore your own amateur industrial design efforts will not compromise the engineering quality of your design, while many of his/her early professional suggestions would do so if adopted.

1. Use your company's trade names, logo, design style and decorations, colors, nameplate design, and other conventions which identify the product as coming from your company. Molded or cast enclosures can provide many of these features at no extra part cost and at only a small increase in tooling cost.

2. Use standard symbols in labels. Use real English sentences in signs. A favorite freeway sign in San Diego reads: "Cruise Ships, Use Laurel Street Exit" (Laurel Street leads to the docks).

3. Make appearance match function. A bulldozer and a sports car should not look alike.

4. Covers should protect the product's insides from the outside world and protect the outside world from danger and contamination from the product's insides. They should mask irregular shapes. They should be easily opened and reclosed but not easily misplaced. Some should have interlocks for safety.

5. External surfaces and enclosures made by molding or casting look far better than those made of bent sheet metal. Such moldings or castings can have functional features included which would have to be built up of many parts if "cheap" sheet metal construction is used—a two-way winner.

This chapter has condensed a major subject into a brief survey, but if this is your first exposure to the subject, the chapter will have been worthwhile.

CHAPTER 24

Accuracy, Adjustment, and Gauging

24.1 GENERAL

Finished products, their subassemblies, and their detail parts must all be made with specified accuracy so the parts and subassemblies will go together and the finished product will work.

Accuracy can be achieved by either manufacturing the parts accurately in the first place or by adjusting parts, subassemblies, and final assembly to the necessary accuracies. Both approaches are used and the choice is not obvious; it is up to your judgment as a designer.

Making the parts sufficiently accurate that they fit together and work right without adjustment is easy to say but may require expensive manufacture. And "fit together and work right" also applies to replacement of parts by field maintenance people. Furthermore if an accurate part wears, it must be replaced, but if there is adjustment, the only maintenance need may be for readjustment.

Whatever the tolerances to which you design the parts and the adjustments you choose to make, it will be necessary to test or gauge the final product. Furthermore, it will probably be desirable to first gauge at least some of the parts and subassemblies and reject the bad ones to prevent costly disassembly of the final product if it does not work.

To assemble a product which does not work costs one unit of labor; to disassemble it costs a second unit; and to reassemble it costs a third. This sequence is a *triple*. (There was a manufacturing manager at MOBOT who committed triples, sextuples, and worse. We should have fired him a long time ago.)

This is not a problem you should buck off to the manufacturing engineers. You can make the cost of gauges and gauging higher or lower by providing gauging features, fixturing surfaces, and so forth as part of your design.

The same applies to field maintenance; if it is easy to gauge the performance of your product, it will cost less to maintain it and it is more probable that it will really be maintained. Field gauging includes both the use of instruments applied by maintenance personnel and the inclusion in the product of meters, indicators, and alarms which call for maintenance.

The subjects of accurate manufacture and gauging belong under manufacturing, but the techniques of adjustment require separate treatment.

24.2 ADJUSTMENT TECHNIQUES

The following techniques are practiced. Some are coarse, some are fine, some can be either, and some require more or less skill than others.

1. Bending. Bending can be as fine as adjusting the contact springs of a small relay with special tools to an accuracy of a few thousandths of an inch, or it can be as coarse as sledging a hot steel plate into shape in a ship's hull. If you choose to use bending, be sure that the material will accept a cold worked set and not drift too much afterward. Also be sure that adequate skill is available both in the factory and in the field.

Adjustment is usually accompanied by some gauging process, which may be only visual examination, to indicate when the adjustment is correct.

An advantage of bending compared to other techniques is that manufacturing cost—other than the bending cost—is minimum.

2. Inclined plane. Inclined plane adjustment elements include:
   a) Screw adjustments have fine resolution, large range, are stable, and require little skill. They require means to prevent the screw itself from drifting. Such means include:
   Friction locknuts (plastic insert, deformed threads)
   Set screws between male and female threaded parts
   Adhesive thread locks
   Castellated nuts and cotter pins
   b) Wedges usually depend on friction to retain their position.
   c) Cams may require a clamp for their position, or they may have a plurality of positive stops analogous to the castellated nut.

3. Slide and clamp. These words are self-descriptive of the technique.

4. Shim. Shims are tedious to install, but they provide high strength to oppose a disturbing force and they are extremely stable.

5. Selective assembly. The best example of selective assembly is in the manufacture of precision ball bearings. Balls and races are graded and sorted by size, and matched sets are then assembled.

6. Cut. Material is removed for adjustment by the following processes:
- Scrape
- File
- Ream
- Grind
- Lap
- Hone
- Wear-in

Some high-accuracy parts such as lenses, telescope mirrors, and extremely accurate lead screws use wear-in. They are finished in pairs with a negative part as a tool. Fine abrasion between the positive and the negative adjusts both to final accuracy.

Remember not to require cutting through surfaces with protective finishes. Remember that cutting is irreversible. Our highly skilled model maker and the author's unofficial teacher in his first job as an engineer used to joke, "I can't understand it; I cut a piece off and it's still too short."

7. Peen. Peening is cold working of a malleable surface by hammering. It is used to straighten bent shafts by causing one side to expand axially.

8. Flame spray. Flame spray is a method of building up a surface by spraying molten particles, usually metal, onto it. It is usually followed by finish machining.

9. Electro-plate. Electro-plating is a method of building up a surface with fine resolution.

10. Vary the geometry. For example, vary the mechanical advantage of a lever. A weighing beam with a sliding weight does this.

11. Discrete steps. A pin or screw or spring engaging one of a series of holes maintains the adjustment established by that hole. For example, a castellated nut and cotter pin.

12. Automatic adjusting machines. The adjustment (perhaps a calibration), if mechanically simple like a screw, can be performed by a servo-motor in a feedback control. A gauging system feeds back the error to the motor. The motor may turn a screw, or it may drill a part to balance it.

This is by no means an exhaustive list of adjustment techniques. You may find others in the literature, and you may invent your own.

CHAPTER 25

Reliability, Defects, Abuse, Failure, and Maintenance

This is a doleful sounding chapter, but your career may hinge on it.

You want the product you designed to work when your customer wants it to work; that is *reliability*. Opposing you are *defects* in manufacture and *deterioration* and *abuses* which can result in *failure* to work; supporting you is *preventive maintenance* to prevent failure and *corrective maintenance* to overcome failures.

What we are talking about are sources of *un*reliability. If your product is made without defects, and if it is not abused, and if it receives proper routine (preventive) maintenance, it will never fail before its life expectancy and its reliability will be perfect.

You, as designer, have a great deal of control over the reliability of the product because abuse resistance, manufacturing, and maintenance strategies are parts of the design.

25.1 DEFECTS IN MANUFACTURE

Defects can enter your product in purchased materials, in purchased components, in manufacturing processes in your own plant, and in processes in contractors' plants. You, as designer, can help specify reliable sources, and your drawings can specify manufacturing processes and inspection procedures.

Some test specifications must be statistical results of testing to destruction so that the actual materials and components used are not tested. Some such tests are under stresses higher than in service, to indicate margin of reliability. Explosive charges, shear pins, and electrical fuses are obvious examples, but electronic components are also in this category.

The result may be expressed as mean time before failure (MTBF) and the reliability of an assembly computed from the MTBFs of its components.

Inspection (gauging and testing) of components and subassemblies during manufacturing may detect defects that, if uninspected, may hide, latent, in final assemblies. You can design your components, and your division into subassemblies, to make such inspections either difficult or easy. You can make your tolerance scheme either expensive or inexpensive to manufacture and to measure. You can make your documentation either easy to understand or easy to misunderstand.

Among the most reliably made products are space hardware, which cannot be maintained other than by orders radioed from the earth, and implanted medical devices, such as pacemakers and heart valves. The soldering procedures for space electronics (NASA soldering) are agonizingly detailed, and the technicians are rigorously trained.

(The author once watched technicians assemble the extremely intricate and miniaturized life detection experiment for the Mars lander. They worked in pairs, a worker and an inspector. The inspector never touched anything, but he never took his eyes off the worker's hands. Not cheap. On the other hand, the author saw that the technicians were using their own common and beat-up tools they used for work at home, although the assemblies they worked on were extremely small. When he said "special tools" to his friend, who was his guide, the friend smote his head, then said it was too late to change! The moral: Your product design may be radically improved if it includes the specification or design of special tooling.)

Material handling and storage during manufacture and shipment can introduce defects. Such *mis*handling can include most of the abuses listed below.

Your design can provide reliability despite some degree and some combination of defects and abuses. For example:

• You can provide a margin of safety in the strength and rating of parts and components.

• You can provide redundancy so that if one part fails, another part takes over its function.

• You can provide thread retention for all threaded fasteners.

• You can provide indicators of imminent failure. Among such indicators are built-in instruments and signals such as some on the dashboard of a car. You can provide reserves of consumables, such as fuel, for example the Volkswagen spare gas tank.

• You can design for *fail-safe* so that a failure may stop further functioning but will not cause outside damage. For example, an irreversible worm gear in a power transmission will prevent the load from moving if the motor fails.

• You can design for *fail-soft* so that a defect will materialize gradually and show gradual deterioration rather than sudden failure. For example a ductile part may deform before total failure, while a brittle part will fail totally without warning. (However, a ductile part *may* fail from fatigue just as totally and without warning as a brittle part.)

• You can provide a package of spare parts, containers of consumables (for example, lubricants), and tools as part of the product or as an extra-cost maintenance kit.

• You can make your product tamper resistant with locks and with a need for special-access tools. The common fire hydrant's five-sided head on its valve stem is a perfect example.

• You can provide adequate and *clear* maintenance warnings and instructions, including labels on the product itself.

Everything costs, so you always face the challenge of designing the best product for the least cost.

25.2 DETERIORATION

Products that leave the factory operating perfectly within tight specifications may begin to lose their performance quality over time. Why? The following is a devil's menu of how to make a product fail. You should review it periodically to make sure that you have made your design resistant or that you have at least warned the user.

1. *Abuse:* Impact, overload, use for other tasks than intended, and so forth.

2. *Heat:* In the sun. In an enclosure, such as a car, or warehouse, in the sun. Near a furnace or engine. In a cabinet with electrical power dissipation. Hydraulic system heating. Improperly used soldering irons and torches during manufacture or maintenance.

3. *Cold:* Lubricants congeal, parts contract differentially, fuels volatilize slowly.

4. *Wet:* Rain and melted snow and sleet penetrate unsealed joints and enclosures and do a variety of mischief inside. Temperature cycles breathe in humid air from which water condenses and remains.

5. *Corrosion:* Seawater. Salt air. Humid air. Fresh (more or less) water. Chemical liquids, gases and vapors, including smog and other pollutants. Dissimilar metals in contact causing electrolytic corrosion. Chemical concentration cells. Scale deposition from water and chemicals. Human body chemicals, both on the skin and internally, are incredibly corrosive.

Implanted prosthetics must be made of a very limited list of materials, including titanium, Elgilloy, vitallium, and some plastics. Some of this limitation is to protect the part and some is to protect the patient. Beryllium copper would make a wonderfully fatigue-resistant spring and conductor if it were not also a deadly poison. Even silicone has now been proscribed.

6. *Dust and dirt:* Penetrating. Coating (sometimes bonded by oil, which oxidizes and resists cleaning). Some are chemically inert; some are chemically active.

7. *Organisms:* Fungus, particularly on electronics. Insects. Larger animals, including rodents. Fish bite on undersea devices. Tampering by humans. If your product is medical, it must not harbor microorganisms and must withstand sterilization.

8. *Shock:* Mostly from dropping, but from other impacts as well, including impact from tools and from vehicle collisions. Military products are subject to cannon recoil shock and to nearby explosions.

9. *Vibration:* During transportation. Contact with vibrating machinery during operation.

10. *Sabotage:* No, not by enemy spies and saboteurs but by disgruntled employees, burglars, rioters, or curious or malicious tamperers. Vandalism.

11. *Overload by operator:* Overload can be the result of carelessness but can also be the result of an operator trying to accomplish too much too fast.

A most egregious and frequent overload abuse is of the overhead baggage compartments in airplanes, by passengers. There is no limit but muscular strength to how big and how heavy the load is which is stuffed into these lightweight plastic boxes and compressed in by shoving the door. Yet there never seems to be a box burst or a door lock break open. A superb piece of design engineering.

12. *Liquid accidents:* Spilled coffee and sugary soft drinks may sound funny, but can clog mechanisms and short or leak circuits. Consider a computer keyboard. Being left out in the weather, overzealous cleaning, and dripping oil or water can do the same.

13. *Fire*

14. *Smoke*

15. *Negligent maintenance:*
- Wrong materials (for example, lubricants)
- Skipped schedule
- Neglected items
- Improperly obeyed instructions (measurement, adjustment, and so forth)
- Tinkering
- Loose fastenings

16. *Evasion of safety devices:* Operators devise ingenious ways to bypass safety devices in order to improve their performance or to reduce their effort. Sometimes a machine is damaged and sometimes the operator is damaged.

17. *Normal wear and material aging*

18. *Looseness and adjustment drift*

25.3 FAILURE MODES

Every electrical device is a mechanical or chemical device which carries electricity, so every electrical failure is, or causes, a mechanical or chemical failure, sometimes deep inside. Common failures include:

1. Mechanical fracture or deformation

2. Mechanical wear-out

3. Fastener loosening

4. Corrosion, including electrolytic from electric current

5. Electrical burnout, open circuit, leak, or short circuit

6. Fluid leaks or clogs

25.4 REPAIR STRATEGIES

You should design your product to be repaired *(corrective maintenance)* in accordance with one of the following strategies. Basic to all of them is the question of which organization does the work: the customer, the manufacturer, the distributor, or a separate maintenance company.

25.4.1 Discard and Replace

This policy makes your design easiest and cheapest. For example, electric lightbulbs are discarded when their filaments burn out, although they could be designed so that one could open them, replace the failed filament, reseal, and evacuate. Automobile and home appliance subassemblies such as small valves and pumps are treated in this manner.

25.4.2 Return for Repair

This is the practice with most portable consumer products. The maintenance depot may be operated by the manufacturer or as an independent business.

25.4.3 Call in a Maintainer

Heavy consumer and commercial products. Production machinery and office machines waste money every minute they are down, so frequent need of repair damages your company's reputation. Factory maintenance branches and maintenance companies provide quick response. At least one major computer company bases its reputation largely on its quick and certain maintenance.

25.4.4 Module Replacement: Return, Repair, and Inventory

The part or assembly is replaced from inventory, returned to the factory or maintenance depot for repair, then restored to maintenance inventory. This policy conserves the initial cost, but it requires that the design make the product capable of diagnosis, disassembly, and repair and that a channel be maintained for flow of material, orders, and money. It is quite popular for electronic printed circuit board assemblies where the diagnosis may be difficult but the replacement part may cost only a small fraction of the board assembly. It is also used for relatively expensive mechanical and EM assemblies. The customer must be willing to accept a rebuilt assembly.

25.4.5 Repair in Place

This policy requires maximum skill in the maintainer, both in diagnosis and repair. It requires that the design permit disassembly and reassembly in the field, usually with only hand tools and often with only standard hand tools. In some ways, this is the most expensive design, but it has the advantage of not requiring the product to be divided into subassemblies separable in the field.

25.4.6 Multiechelon Maintenance

Used mostly in the military, this policy divides maintenance into:

• module replacement or adjustment by a user or first-echelon technician close to the location of use;

• maintenance within those modules by a more skilled, second-echelon technician at a maintenance depot; and most subdivided maintenance by return to the factory.

25.5 ROUTINE MAINTENANCE

One of your design choices is either to establish a life period for a part or subassembly and order it to be changed on schedule or else to permit the part or assembly to run until it fails, which may be the entire service life of the product. The first choice is more costly in maintenance, but may increase the reliability of the product by preventing random failures. Of course, some items must be replaced on schedule, such as filters and fluids in automobiles.

As with any other human activity, routine maintenance can be done well or badly. For example, there is in some companies an opinion that pneumatic devices are less reliable than hydraulic devices. Not true. The explanation of this paradox is:

Shop air is wet, dirty, and poorly regulated for pressure. The air for pneumatic devices must pass through a filter, pressure regulator, and lubricator (FRL) usually provided as part of the air-powered device. If the filter is not cleaned, or the pressure drifts, or the humidity is very high, or the oil is not replenished, air cylinders, valves, and motors *will continue to work anyway* but will gradually wear out and become erratic—they fail soft. Therefore these maintenance functions can be neglected without immediate consequences. But everyone knows that dirty oil will instantly jam hydraulic systems, so hydraulic filters are carefully tended and their systems are reliable!

Prejudice makes for self-fulfilling prophecies. If a user believes that electronics, or pneumatics, or hydraulics, or relays, or particular devices or machines are unreliable, they will be. Consciously or unconsciously, the user will make the prophecy come true.

25.6 DESIGN FOR CORRECTIVE MAINTENANCE

One thing you can do to improve maintenance is to make it easy. Accessibility is the watchword. For example:

1. Provide access doors.

2. Provide access from ground level (without a ladder) and from as comfortable a position as you can.

3. Do not jam things into tightly fitting spaces. (There are designers who shove a bunch of electrical connections into a small hole with sharp edges.)

4. Require a minimum of tools, preferably standard. On the other hand, a special tool can be a blessing to a technician; this is yet another call on your judgment.

5. Make it easy to disconnect and *correctly* reconnect hoses, pipes, and wires. Make it impossible, or at least difficult, or at least obvious, to reconnect improperly.

6. Make it easy to replace belts and other power transmission components which can wear out.

7. Make it easy to remove and replace fasteners. Do not require a hand to reach around an obstacle to hold a nut, or three hands to make an assembly. Remember that when replacing a part, it is tempting to omit washers, lockwashers, and even some screws.

8. Use standard fasteners with a minimum number of sizes so that a lost fastener can be replaced easily. If a special fastener is needed, make it very special so a standard fastener is *not* used as a replacement.

9. Use a minimum number of screw head varieties; hex and socket heads are the easiest to use. Use thread-retention fasteners. Remember that a thread adhesive used in the factory will probably be omitted in the field.

10. Use screws and nuts rather than screws and tapped holes. A stripped thread in a tapped hole calls for a major repair effort. Furthermore nuts are cheaper than tapped holes and are less likely to have manufacturing defects.

11. Build in fluid and electrical ports for easy diagnosis. Build in windows and indicators for the same reason.

12. Build in running time meters and alarms to encourage correct routine maintenance.

13. Use standard fluids whenever possible. Automobile lubricants and other fluids are readily available almost everywhere.

14. Use easily available standard components. This is another judgment call; special components may be worth their trouble.

15. Predict field or recall retrofits for possible mass failures, and provide means to make them. This is not like predicting the stock market; if you know your product and your industry, you may be able to make some shrewd guesses about the future. (See number 22, below.)

The things you do to ease field maintenance will also reduce manufacturing labor; they are *two-way winners.*

16. Plan a spares program and offer spares kits and ordering information.

17. Document installation, operation, and maintenance instructions. Include theory of operation, assembly and subassembly drawings, preferably exploded with callouts, and parts list. A complicated device should have a diagnostic tree. Even if your company provides the luxury of technical writers, documentation is part of design; it is your professional responsibility, as is writing it in easily understood and unambiguous English. Hate it, but do it.

18. Document installation instructions. Much harm can be done by improvising installation procedures. Design the product to be improvisation-resistant, although this is much easier said than done. Remember the rule, "When all else fails, read the manual."

19. Put warning signs on the product. "High Voltage," "No Step," and so forth.

20. Electric power switches are available with padlock holes so a maintenance worker can lock off the power with a personal padlock. In a large machine, it is possible for worker A to turn on the power without realizing that worker B is exposed.

21. Provide model numbers and serial numbers on both the product and its components. You may have to trace an operating problem or a failure to its source, and these numbers may help you. You may have to have a recall for a defect discovered after shipment, and these numbers may be essential. For example, the newspapers frequently recite automobile recalls. (See number 15, above.)

22. Your product will bear some warranty, if only the legal one in the Uniform Commercial Code. Be sure that warranty is justified. Nowadays, not just the company, but an individual design engineer—you!—is legally liable for damages from negligence in designing a product which causes harm. Keep a log describing the evolution of the design, particularly the detection, diagnosis, and cure of defects.

AS AN ENGINEER YOU MUST ALWAYS BATTLE MURPHY'S LAW: "IF SOMETHING CAN GO WRONG, IT WILL." PHILOSOPHICALLY MURPHY'S LAW IS AN ASPECT OF THE LAW OF ENTROPY, BUT IGNORANCE OF THE LAW IS NO EXCUSE.

CHAPTER 26

Barriers, Filters, Conduits, and Valves

26.1 GENERAL

A substantial portion of design is blocking flow with barriers, or selectively blocking flow with filters, or guiding flow with conduits, or adjusting flow with valves. Flow of what? Here is a list:

- Gas
- Particle cloud in gas
- Vapor
- Liquid
- Slurry
- Electricity
- Magnetism
- Microorganisms
- Animals

- Humans
- Force
- Noise
- Heat
- Information
- Dirt
- Privacy
- Ugliness

26.2 EXAMPLE: THE CAR

You might think that some of these belong only to chemical or electrical engineers and you need not bother. Right?

Wrong. Consider a car. Certainly an EMD. Yet in it:

You provide *barriers* against leakage of oil, gasoline, brake fluid, transmission fluid, coolant, bearing grease, combustion gases, windshield wiper fluid, battery acid, battery and generator voltage, ignition voltage, electrical noise radiation, some exhaust pollutants, and exhaust noise.

For the engine, you provide *filters* for gasoline, oil, and combustion air. You provide a catalytic converter for the exhaust gas (It is not exactly a filter but is a chemical processor which filters out smog-producing components of the exhaust gas by oxidizing them before they are exhausted).

For the car's passenger compartment, you provide air which is dirt-*filtered,* heated, or cooled.

You provide locks to the passenger compartment, the ignition, the glove compartment, and the trunk as *filters* to pass certain humans and not others.

You conduct *(conduit)* information to the driver via mirrors and dashboard instruments. (More and more these instruments are electronic, but electronic information systems start with EM transducers to convert mechanical information to electrical information.)

You may tint the windows for privacy *barriers* when requested to do so. In some cars, you also provide privacy curtains on the windows and a soundproof window between the passenger and chauffeur compartments. In many taxis and police cars, you make this barrier criminal resistant but speech conducting, so it becomes a filter. Privacy glass and shades may totally block vision, or dim images, or distort images without concealing their presence.

You provide the passenger compartment with noise *barrier* insulation and both the passengers and the outside world with an engine noise *barrier,* the muffler.

You paint and plate surfaces as *barriers* to corrosive gases and vapors, also to the sight of ugly surfaces. You provide a variety of decorative covers as ugliness *barriers.* (Ugliness and beauty were discussed in Chapter 24, "User-Friendly Design.")

You provide heat flow *conduits* out of the engine into the atmosphere and *conduits* and valves for heat flow into and out of the passenger compartment.

26.3 BARRIERS

26.3.1 Fluids

26.3.1.1 Static seals

To prevent leakage of gas, vapor, liquid, or slurry through a joint between parts which do not move relative to each other, you use O-rings or gaskets. O-rings require little clamping force and little width, but the mating parts must be accurately machined and the elastomeric materials used have relatively narrow temperature limits. Metal O-rings exist which have very wide temperature limits.

(The *Challenger* disaster was due to an ignorant design error in the width of an O-ring groove. The error should have been caught by the drafter's checker, and certainly by the seals engineer. (See *Real-World Engineering* [A3] pp. 5–7.)

Seals between coaxial cylindrical surfaces cannot be directly clamped against the surfaces and therefore must be either O-rings or expanding packings.

Gaskets require high clamping force and substantial width but need less accurate machining than for O-rings. Some gasket materials have very wide temperature limits.

Among static barriers are anticorrosion, antifungus, and antimarine life coatings. They include paint, varnish, electro-plated or flame-sprayed metal, organic material wrapping, and chemical conversions of the surface. Since some corrosion is electro-chemical in nature, some coatings are sacrificial; that is, the coating corrodes away instead of the underlying structure. Zinc coating on iron *(galvanizing)* is a common example. Steel boat hulls may have actual blocks of zinc for this purpose.

Steel pipelines may have electric current sources attached to prevent electrolytic corrosion by providing current in the reverse direction from the corrosion current.

Antiorganism coatings such as boat bottom paints and electrical printed circuit board varnishes often contain poisons to inhibit organism growth.

No coatings have yet been accepted to seal stone sculptures and buildings from corrosion by smog and salt air; old monuments continue to decay. Nor is protection from bird droppings even mentioned.

Flexible hose connection static seals usually use a shrinking or expanding ring between the hose and the mating rigid member. Often the hose itself is compressed radially inward against a rigid tube.

Metal tubing is sealed either with a variety of metal cones, sometime formed into the ends of the tubing, under axial pressure, or with O-rings or compressed packings.

An electrical cable entering a box is sealed with an elastomeric sleeve which is axially compressed and bulges inward against the cable and outward against the box fitting. (It took two years to persuade a client, who made highly sophisticated electronic cardiac pacemakers, to even try this EM classic! It worked fine.)

A perfect fluid-tight seal for an electrical conductor which passes through a hole in metal is an annular glass bead which is bonded to both the conductor and the hole. Both conductor and hole are of metals with coefficients of thermal expansion close to that of the glass. Typically, a commercial assembly comprises a metal plate with one or more holes, glass beads, and short conductors. The plate is soldered into the box and wires are soldered to both ends of the short conductors.

Pipe is sealed by tapered threads in which is a hardening compound or conforming gasket cord. Welding is the ultimate seal for pipe. (Unless the weld is porous! Many welds are X-rayed to detect porosity.)

26.3.1.2 Rotary seals

Rotating shafts and tubes often must be sealed to stationary structures. For example, your rotating car crankshaft must be sealed to its engine where it exits to the transmission in order to prevent oil leakage. Some rotary seals are for continuous rotation, as in that engine, and some are for intermittent rotation, as for a valve stem. Many valve seals are for a combination of rotary and linear motion. There may or may not be a substantial pressure difference between the two sides of the seal. Among the seals used are flat-face seals, O-rings, and axially compressed packings. The sealed fluid may provide lubrication.

26.3.1.3 Sliding seals

Pistons and steam engine reciprocating valve stems are examples of sliding elements which must be sealed to stationary members. Seals include split piston rings which permit some blowby, O-rings, V-rings, and U-rings.

Some seals provide a substitute fluid which leaks in preference to the sealed fluid. A philosophical relative of this technique of adding something to block something else is the addition of "white noise" to space occupied by people to mask disturbing sounds such as other people's conversations.

Some seals use magnetic particles in oil to bridge the gap in the manner of a magnetic particle clutch.

A zero-leakage sealed drive is a magnetic-driven member inside a thin non-magnetic shell and driven with a moving magnetic field generated outside the shell and passing through it, Fig. 3-4. "Canned" electric motors use this principle.

A different zero-leakage sealed drive is a driven member inside a thin, metal shell which is flexible, such as a bellows, and driven by flexing the shell.

Two applications for such zero-leakage drive seals are sealing radioactive fluids and sealing mechanisms such as switchgear inside high vacuum.

The rotating seal problem is finessed in an electric refrigerator by enclosing the motor inside the sealed space and providing static seals for its lead wires.

26.3.2 Mechanical Oscillation

26.3.2.1 The spectrum

Mechanical oscillation occurs in a broad range of frequencies. Since an adult can hear oscillation between approximately 100 cycles per second and 15,000 cycles per second, we call that range *audible sound.* Below the audible sound range, we call oscillation *vibration,* and above the sound range we call oscillation *ultrasound* or *ultrasonics*. Vibration, audible sound, and ultrasound are exactly the same thing except for frequency.

(Until approximately 1946, we measured frequency in *cycles per second* or just *cycles* for short and got along just fine. After World War II, we recruited German engineers and scientists who had developed Hitler's V1 and V2 weapons and brought them to the United States to work on our own rocket programs. The recruiting drive was code named "Operation Paperclip," and we competed with the Russians who were doing the same thing.

The Germans found that although most electrical units were named after famous scientists (volt after the Italian scientist Volta, ampere after the French scientist, Ampère, and so forth), oscillation frequency was named in simple English. Opportunity!

The Germans declared that cycles per second should be called *hertz* after the famous *German* scientist Hertz, and they sold the idea. So now we refer to the above frequencies as 100 *hertz* (or 100 Hz) and 15,000 *hertz* (or 15,000 Hz). Pure prestige wordsmanship with zero new information.)

26.3.2.2 Vibration

We provide barriers to oscillation both by damping it at the source so there isn't so much, by filtering out whatever we can, and by insulating it between the source and anything it might bother, such as our bodies, our ears, and our machines and devices. For example:

An automobile wheel, rolling along a road that is less than glassy smooth and flat, is oscillated up and down by the road, and the rougher the road and the faster the car, the rougher is the oscillation. First-stage damping and filtering is by the tire itself. The rubber is both a spring and a damper, and the air in the tire is also a spring. The oscillation which gets through the tire to the wheel is damped by the "shock absorber" (which it is not—it is a damper) and further filtered by the suspension spring. The oscillation which reaches the chassis is further filtered and damped by the springs and padding in the seats. (All of this is subject to mathematical analysis in the *science* of design.)

26.3.2.3 Audible sound

Vibration does not reach us through the air, but audible sound does. A variety of insulations are barriers to sound transmission to a human space and reduce reflections within a human space. This art is part of the field of *acoustics*. One of the problems of mechanical design is to prevent structural parts from conducting sound, bypassing insulation.

A special case is noisy machines such as internal combustion engines and air-powered tools, which exhaust gases, and office machines that require cooling air. A filter is required that passes exhaust gas but blocks exhaust noise. Such filters are called mufflers. They use combinations of labyrinths, resonant chambers, expanding pathways and sound-absorbing materials.

26.3.2.4 Ultrasound

Ultrasound is conducted by the air, although humans do not hear it. The burners of a hot air balloon generate enough ultrasound to drive dogs on the ground into a frenzy. And, of course, there are ultrasonic dog whistles.

Ultrasound is severely damped by most materials, so there is not much of a problem in providing barriers. In fact, the principal mechanical design problems involving ultrasound are to provide conduits for it so that damping it can provide heat in thermoplastic workpieces.

26.3.3 Electro-Magnetic Shielding

Electro-magnetic radiation is a spectrum which extends from long radio waves through heat waves, visible light waves, X-rays, to gamma rays. All radiation follows exactly the same physical laws, but practical dimensions for devices and the different absorption and reflection behavior of different materials result in different conduits and barriers for different wave lengths.

All electro-magnetic radiation is transported exactly the same way through empty space. The atmosphere behaves almost like empty space except that particles and vapors selectively absorb radiation differently at different wavelengths. (During the development of radar, there was a crisis when it was found that water vapor in the atmosphere selectively absorbed certain of the wavelengths chosen.) Heat radiates through dust-free air almost as well as through empty space.

The usual barrier to long waves is a surface of conductive material, either solid or mesh. The conductive surface is a short circuit to the electrical component of radiation.

A sheet of high-permeability magnetic alloy is a barrier to both the magnetic component of radiation and to DC and AC magnetic fields.

Heat radiation is discussed in the next section.

X-radiation and gamma radiation are attenuated by heavy metals such as lead. The thicker the layer of barrier metal, the greater the attenuation.

Nuclear fission particles are slowed and absorbed by heavy metals. Neutrons are slowed by heavy hydrogen (in water) and by carbon.

26.4 HEAT

Vacuum, as in thermos and Dewar bottles, is a perfect insulator for conductive and convective heat flow (and for sound) but offers no resistance at all to radiation flow. Usually, the vacuum chamber walls are covered with a reflective coating to reduce radiation transfer, but the vacuum is not perfect and the reflectivity is not perfect. For hot coffee and laboratory gases, the net effect is quite satisfactory but when it became necessary to store large quantities of liquid oxygen and liquid hydrogen for large rockets, the insulation was inadequate that is, too much liquid boiled away. *Superinsulations* were developed to fill the vacuum space. They used either fine laminations of reflective foil spaced by insulating fibers or fine powders of insulating ceramic.

Heat conduction is resisted by non-metallic materials such as plastics, fiberglass, and so forth. However there is no perfect barrier to heat flow, sooner or later heat will get in or get out.

26.5 FILTERS

26.5.1 Particles

Particles may be filtered from gases and liquids by passing the gas or liquid through channels smaller than the particles. A common structure is a porous body in which the pores are smaller than the particles to be blocked. The porous body may have very large pores, such as the holes in a wire screen, medium-size pores, such as the holes in filter paper, or extremely small pores such as the holes in the membranes used in reverse osmosis desalination, which block sodium and chlorine ions. When the pores become blocked, there is a variety of techniques for unblocking, such as reverse flow or by simply replacing the porous body.

Centrifuges provide filtering by throwing denser particles farther out than lighter particles. Centrifuges are used to separate uranium isotopes where the particles are single molecules which differ by only the difference in atomic weight of the isotopes. A form of industrial centrifuge is called a cyclone. Incoming gas is deflected into a helical path inside a shell, dust particles centrifuge to the outside of the shell, and clean gas is drained from the inside.

A gas centrifuge called a vortex tube is similar to a cyclone, but separates high-velocity molecules from low-velocity molecules, which are colder, and supplies the colder molecules as a source of refrigeration.

A settling tank is similar to a centrifuge, but operates on only 1G whereas a centrifuge may generate thousands of G.

One of the best filters ever invented for gas-borne particles is the EM Cottrell precipitator. The gas passes through a grid of fine wires from which a high-voltage corona charges the particles. The gas then passes between a set of parallel plates with alternate plates connected to high voltage and to ground. The electrically charged particles are attracted to the plates and settle on them. The plates are periodically cleaned by rapping or by washing. Cottrell precipitators are made in very large sizes used to remove smoke and fly ash from coal- and oil-burning power plants and in small sizes used in homes. They remove particles as small as cigarette smoke, but unfortunately, they have no effect on sulphur dioxide molecules, so we still have acid rain.

Clean rooms and glove boxes use air from which fine particles have been filtered. Clean rooms are largely self-defeating because the people inside continually foul them with hairs and breath vapor and droplets. Face masks are commonly pulled down below the chin to make breathing easier, exposing the mouth and nose.

Glove boxes are smaller, cheaper to buy and to operate, and cleaner than clean rooms. They also permit using inert or other chemically desirable atmospheres in which people cannot breathe because we require oxygen. Machines can have lubricated and other contaminating parts external to the box with working parts extending into the box through seals or bellows. A spot welder for a glove box in a semiconductor plant works on this principle. The only disadvantage is the gloves' impediment to dexterity, but proper tools can compensate this. No special clothing is needed.

26.5.2 Animals

Animal filters are fences, gates, and differential obstacles. Cattle are filtered out by rows of bars in the pavement over which they cannot walk but which are close enough to support auto and truck wheels and can be easily crossed by humans. Stiles are stair filters which can be passed by people and dogs but not by cattle and horses.

26.5.3 People

People filtering, that is, *security,* is a major design problem. Walls, fences, locked gates, window bars, human guards, and guard dogs are barriers which try to prevent unauthorized human passage. Lock keys, combination lock codes, passwords, badges, and other identification schemes permit authorized humans to pass. Lights and alarm systems inhibit some unauthorized humans and call police to compensate for unauthorized passage.

In the civilian world, people traps are illegal. But in the military regime, land and sea mines and booby traps inhibit passage of humans who do not know their locations but permit passage of people who do. (There are now millions of land mines left in the wakes of wars. They blow the legs off children and adults every single day. Please consider the development of mine-clearance technology a standing challenge to every one of you.)

A friendly kind of people filtering is the safety guarding of machinery. Belts, chains, gears, and conveyors are all traps for the unwary. In machine tools and presses, human hands must enter a space to load and unload a workpiece, but that same space is occupied with harmful or lethal portions of the machine when it is its turn to manufacture the workpiece. A great variety of mechanical, electrical, optical, and EM safety devices have been devised, and many are in use. Unfortunately, they tend to slow down the rate of production. The machine operator who is being protected is motivated to evade that protection by the virtuous desire to increase productivity, and intends to be careful.

(The president of a plastic molding press company told of a worker who evaded four concentric safeties by reaching over the top of the guard fences, and lost a hand. The president wanted to equip his presses with robots to eliminate the workers, their ingenuity and their risked hands, but his chief engineer and sales manager told him that the customers would not pay for them.)

26.5.4 People Guarding

Filters and barriers worn by people to protect them include sound-insulating ear covers; light-dimming welding masks and laser light masks; heat-resistant clothing; rubber gloves for electrical and medical work; helmets, masks, safety goggles, and steel-toed shoes for impact protection; and chemical-resistant clothing.

The police and the military use chemical- and bullet-resisting clothing, helmets, shields, and masks.

At a more mundane level, gloves, warm clothing, shoes and boots, raincoats and umbrellas are environment barriers for people; lightweight clothing are modesty barriers to light. Condoms need no further comment. This book will not try to fit fashion into this scheme.

Although not really filters, snorkels and scuba gear enable people to breathe under water, and wet suits are warm clothing for cold water. Firefighters have breathing gear similar to scuba to protect them from inhaling smoke, and soldiers have gas masks and chemical barrier clothing.

In surgery, barriers are sewn in to close holes in the septums of human hearts and are sewn over arteries to close off aneurisms. The surgeon's mask, gown, and gloves are barriers keeping *in* the surgeon's germs.

All of the above are the result of engineering design, although not all the designers were engineers.

26.5.5 Electrical Oscillation

Filtering electrical frequencies is an elaborate, analytical, established art. Most filters are analog electrical circuits, but recent advances include digital filtering by computer.

26.6 CONDUITS

Just to be complete, let us recite that gases, vapors, liquids, and slurries are conducted by pipes, tubes and hoses, and sometimes open channels; that electricity is conducted by solid and liquid metals, ionized liquids, and ionized gases; that heat and vibration are conducted by all solids; and that people are conducted by paths, ladders and steps, elevators and escalators.

Now let us consider some less obvious forms and relationships.

A complex of fluid conduits, or passageways, can be cast or machined into blocks of material. The result is a 3D analogue of a printed circuit etched from a continuous sheet of copper bonded to an insulating substrate. Automobile automatic transmissions have such intricate conduit arrays in their die castings.

Heat is transported by moving fluids, including furnace gases, boiled water (steam), boiled mercury, boiled refrigerants, air (including refrigerated air to be reheated by devices which must be cooled), and the vapor in heat pipes. Some fluid transportation is driven by convection due to the different densities of warm and cold fluids, and some is driven by pumps and blowers. In heat pipes, the liquid is driven to the warm end by capillary action, and the vapor is driven to the cool end by the pressure drop due to condensation.

Powders and solids (such as cash containers) are transported in tubes along which they are driven by compressed air.

Very short-wave radiation may be conducted in round or rectangular waveguides and on microwave stripline. Optical fibers are waveguides for visible light. Telephone, TV, and data communication is increasingly done via light pulses in optical fibers because of the enormously greater rate of information transport which can be carried over an optical fiber than over a wire or a radio channel. (The common word is *fiber optics*.)

All electrical and optical conduits and barriers are mechanical things which conduct or resist the passage of electricity or light.

There is no barrier to the passage of a magnetic field, but iron, nickel, and their alloys provide conduits for it which can bypass a space where the field is unwelcome. *Magnetic shielding.* Diamagnetic materials such as bismuth really do resist magnetism to a tiny degree but with no practical effect except in some instruments. Anyone with a good magnetic insulator should announce it at once.

Analogous statements can be made about electro-static fields.

We usually want a conduit to offer as little resistance to flow as we can get. However to get desired pressure, voltage, or flow distribution in conduit networks, portions of conduit are sometimes made more resistant than the remainder, for example, electrical resistors, and fluid orifices.

26.7 VALVES

We must temporarily close or open conduits either entirely or partially; either by hand, or by remote control, or automatically. The generic word for such a device in a conduit pathway is "valve," but in common parlance, a valve is a device for liquids and gases. (In England, vacuum tubes are called valves.) Let us list a variety of valves including both fluid and other valves so you can see the relationships among them:

1. Electrical valves include relays, switches, circuit breakers, fuses, vacuum tubes, diodes, and transistors.

2. Fluid valves include on-off valves, transfer valves, check valves (fluid diodes), and throttling valves. They may be operated or adjusted by hand; by electrical, pneumatic or hydraulic signals; or by mechanical signals (such as the float-operated valve in a toilet tank). They may be operated by the direction, pressure, or speed of the fluid itself, such as check valves, safety valves, or flow control valves.

3. Most heat valves ar fluid valves for the fluids which carry the heat. Radiant heat valves made as radiation absorbers or reflectors have been made for satellite temperature control. (The author once made a conduction heat valve which bridged two pieces of metal with a variable area of mercury. Worked, but failed to sell it to NASA.)

4. Clutches and brakes are valves for mechanical rotation. The jaw clutch is an on-off valve.

5. In surgery, temporary clamps are used during operations to close off blood vessels. Artificial check valves are implanted inside hearts. Artificial pressure-relief valves are implanted to drain fluid from the skulls of hydrocephalic patients.

6. Doors with locks are valves for people. Doors with latches are valves for animals and babies.

This chapter, in a section on design, does not tell you how to design barriers, filters, conduits, and valves, but it may help your thinking in using them in systems.

CHAPTER 27

Ecology

This chapter discusses the effects on the environment which design engineers can influence by their work. It is not a general text on ecology and does not go into public-policy questions which we can influence only as citizens. The references include a more general work on the subject [A48].

We have always designed resistance *to* the environment into our products, but only recently have we emphasized constraining the products' effects *on* the environment.

You cannot design a device to have zero sensitivity to any environment and zero effect on any environment. Perhaps a platinum crucible. What burdens to impose on the environment, and what to impose on people, in exchange for what benefits to people, are difficult judgments philosophically and economically.

Pollution is regulated by the U.S. Environmental Protection Agency (EPA) and in some cases by the federal and state Occupational Safety and Health Agencies (OSHA). As a designer, it is incumbent on you not to make designs which violate their rules, because they will require expensive redesign when the violations are discovered. The ecology problem gives you yet one more category of helper to consult in your concurrent engineering.

On the other hand, ecology protection generates design projects for you, both in redesigning present polluting products and processes and designing new products specifically to prevent pollution.

Ecology problems fall into these categories:

1. Operating emissions discharged to the atmosphere
2. Operating emissions discharged to the earth

3. Emissions discharged during manufacture of the product
4. Scrap discharged to the earth
5. Consumption of resources during manufacture
6. Consumption of resources during operation

27.1 OPERATING EMISSIONS DISCHARGED TO THE ATMOSPHERE

1. Engine combustion products, producing both smog and poisonous carbon monoxide
2. Fuel evaporation during refueling, producing smog
3. Refrigerant leakage and storage can propellants, reducing the ozone layer
4. Chemical plant and chemical transportation leaks, releasing poison gas
5. Carbon dioxide from fuel burning, causing "greenhouse effect" global warming, with uncertain consequences, probably including the increased melting of polar ice and climate changes
6. Incompletely decomposed incinerator emissions
7. Radioactivity leaks
8. Carcinogens such as asbestos fibers and cigarette smoke
9. Dusts
10. Smoke
11. Odors
12. Heat
13. Acoustic noise. Below a certain level, noise is only a nuisance. (Information bearing sound from neighbors and their sound systems can be a worse nuisance.) Above that level, depending on frequencies, noise can injure hearing. Such noise is most common inside factories and on airports.
14. RF noise (long controlled by the Federal Communications Commission)

27.2 OPERATING EMISSIONS DISCHARGED TO THE EARTH

1. Liquid wastes, such as used lubricants, cleaning liquids, and so forth
2. Radioactive waste, liquid and solid
3. Medical infectious waste

4. Electricity. Ground currents can cause corrosion. Exposed conductors can cause shock and fire. These conditions have long been guarded against by appropriate technology. Electric and magnetic fields from power transmission systems and extra-low-frequency radio transmitters *(ELF)* are now accused of being carcinogenic.

5. Garbage drainage in leaking landfills

6. Nutrients to animal vermin

7. Nutrients to undesirable plants (for example, algae)

8. Fertilizers and pesticides which both help and hinder the beneficial effects of farming

27.3 EMISSIONS DISCHARGED DURING MANUFACTURING

Many manufacturing processes such as electro-plating and etching require, or generate, pollutants. It is the responsibility of the factory to dispose of, or recycle, them but as a designer, you are responsible for requiring them in the first place. This is not to say that your designs should require only benign materials in their manufacture, but if you have a choice, you may want to choose such ecologically favorable designs.

Process chemicals released to the atmosphere (paint solvents, cleaning agents, and other fluids, not at first thought of as "chemicals" by EM engineering designers), *are* chemicals.

27.4 SCRAP DISCHARGED TO THE EARTH

1. Lead, such as from storage batteries

2. Other heavy metals (mercury, and so forth)

3. Organic chemicals such as polyvinyl chloride (PVC) plastics which slowly react chemically to form pollutants

4. Packaging. Even when packaging materials are benign, they occupy garbage space when they are discarded. Furthermore packaging is costly. Recycleability is a benefit. In Germany there is a requirement that some packages be made reuseable and be returned to the factory for reuse.

27.5 CONSUMPTION OF RESOURCES DURING MANUFACTURING

Some manufacturing materials are plentiful, such as steel; some are scarce and getting scarcer, such as tungsten, timber, and petroleum. Some are subject to interdiction by hostile governments. As a designer, you do not want to hurt your company by making it dependent on an uncertain supply.

27.6 CONSUMPTION OF RESOURCES DURING OPERATION

The principal consumable in limited supply is fuel. Petroleum-based fuels have some finite source life before exhaustion. Coal is plentiful but pollutes the atmosphere with sulfur dioxide, causing acid rain.

You may have the opportunity to design equipment to reduce these problems.

27.7 DETRITUS OF WAR

Land mines, unexploded ordnance, defoliants, and other chemical agents are leftovers of war. In northern France, shells from World War I, buried since 1918, still occasionally explode.

27.8 NATURAL POLLUTANTS

To put the sins of technology into perspective, let us consider the many natural pollutants, so-called by human standards. For example:

1. Pollens are allergens to a substantial fraction of the population. (But it would not be such a good idea for agriculture and horticulture to eliminate them, except in indoor spaces with air filters.)

2. Although automobiles brought their emissions, they replaced working horses, and their emissions, and their flies. (If you visit Zermatt in superclean Switzerland, which romantically forbids cars and uses horses, you will get a demonstration.)

3. Bee stings and other insect bites take a substantial toll. In some countries, snake bites kill many. Fishers of natural pearls died from the bites of sea snakes. Divers and swimmers are threatened by sharks and barracuda. There are poisonous plants and fish.

4. Volcanic dust and lava outdestroy any factory's emissions. But nothing is all bad: the archaeologists have Pompeii.

5. Disease germs are the most harmful pollutants in existence. (It is medical *technology* which has radically reduced their destruction.)

CHAPTER 28

Money

"An engineer is a person who can do for one dollar what any darn fool can do for two."

28.1 ITS IMPORTANCE IN DESIGN

Money is an engineering parameter like voltage, force, and energy. Considerations of cost enter into almost all design decisions.

A commercial business pays an engineer to design products which it can sell against its competition at a profit; he/she is a participant in its business, and when he/she took his/her job he/she implicitly committed himself/herself to help make that profit. He/she does this by making the best designs which have the lowest cost.

A not-for-profit organization such as a university or government laboratory has a budget. It is supposed to produce the best and most value it can for that budget, and it pays the engineer to help it do so.

If an engineer is his/her own employer in a business startup he/she will automatically have the same concern for cost that he/she has as an employee, only probably more acutely.

In the real world, you will find that some other employees have a less-dedicated attitude. Sooner or later you will find one who is an outright crook. Sooner or later you will disagree with you employer about your pay and rank. Sooner or later you will encounter a customer or vendor who wants to cheat your organization, perhaps with a bribe. Does any of this reduce your responsibility? There certainly is temptation.

We are concerned here, not with morals, but with professional success as a design engineer and the feeling of satisfaction it brings. For this success and feeling, an engineer must be continually concerned with economy as one measure of the quality of his/her work. Furthermore, he/she will be judged from time to time for raises, promotions, and acceptance in better jobs, and a major part of that judging will be the economic worth of his/her designs.

Do not confuse a desire for economy with greed. Greed is an emotional desire to acquire money. Investors are greedy and they like entrepreneurs who are greedy. But *economy* is a design criterion for the work of the professional designer. As a human being, the designer may or may not also be greedy, but that is independent of his will to produce economical designs.

The remainder of this chapter and the next chapter are devoted to comments and suggestions on economy.

28.2 OVERHEAD

Overhead is an accounting theory word for costs which do not directly contribute to performance of a specific job, for example, electric light and rent. The real-world meaning is, "Overhead is whatever the accountants choose to call overhead, and it is calculated in whatever way they choose."

The usual attitude of management is that overhead is a bad thing and whatever charges can be allocated to direct cost should be. Some government contract rules encourage this. Some anecdotes:

1. At one time in X Aerospace Co. each design room had a magic box. If you rolled up a drawing with a note calling for reproductions and put it into the box, and went back to work, and came back to the box an hour later, you found that the blueprint fairy had made your reproductions and attached them to your drawings.

A clever fellow then invented a way to reduce overhead: He fired all the blueprint fairies from their minimum-wage jobs. Now design engineers walked their drawings from their own building to the repro building, waited in line, waited while the work was done, and walked back. Not at minimum wage. But overhead had been reduced.

2. Long-distance telephone was used to call vendors and customers. In a big company, the phone bill was correspondingly big, and of course, was overhead. Another clever fellow changed the system. Now, to make a long-distance call, you requested it from a company operator, who put it through in an average of 20 minutes and charged the call directly to your job. This successfully reduced overhead. It also discouraged long-distance phone use. But once you had placed the call, you were chained to your desk so you would not miss it when it came. During the wait, you had to switch to a different line of thinking because you needed the call to continue your original line. The consequent increase in direct cost, including the effects of exasperation and reduced information transfer, were never recorded. But overhead was reduced.

The *anyhow theory of accounting* states that "the engineers get paid anyhow, no matter what they do or do not do with their time, but overhead expenses (which might make that time more productive) can be prevented and therefore should be."

The next chapter deals more specifically with design and money.

CHAPTER 29

Cost Reduction and Product Improvement

29.1 VALUE ENGINEERING

At one time, the process of cost reduction after initial design was creatively labeled *Value Engineering,* and some engineers promoted whole departments to second-guess other engineers. It's less irritating if you do it yourself before they get their hands on it.

29.2 CHECKLIST

This is a guide and checklist for hard thinking by you. There are no cookbook formulas. (There are, but most are platitudes. Perhaps some of these items are, also.) These items apply to both initial design, if you have time and budget, and to redesign. When you see the finished product per your initial design and if you get complaints about its cost and function, some of these will stand out. If we were infinitely smart, we would do an unimprovable job the first time.

1. Design for cheaper material.

2. Design for less material, including scrap.

3. Design for automatic part fabrication.

4. Design for automatic assembly.

5. Design for unconventional manufacturing processes, if they provide benefits. We should always follow developments in manufacturing technology.

6. Design for tooled processes instead of for manual work.

7. Design the package and the product as a system.

8. Design for fewer parts. Combine two or more parts into one, including fastening means.

9. Design for fewer fasteners. Provide male and female features on parts so that they self-align. Make them elastically latch together or be inelastically deformed to hold together.

10. Design for assembly without reworking parts to fit.

11. Design for available fabrication and assembly skill. Most factory managers resist the cost and effort to expand their technology and increase the skill of their workers.

12. Choose between a single worker assembling from a kit and many workers on an assembly line. There are benefits and malefits to both.

13. Design different models and products to use common parts, particularly parts already in production for other products. Consult your group technology catalog if you have one.

14. Design similar parts to use the same tooling. Some tooling is adjustable to accommodate a variable dimension. In some tooling, merely omitting drilled holes and the like provides different parts.

15. Design the tooling, at least schematically, together with its part.

16. Specify the loosest tolerances which will not reduce performance. Study the tolerance budget and allocate tolerances in an optimum manner. Avoid "standard tolerances" as expensive laziness. Recognize that a tolerance specification is really a process and inspection specification.

17. Decide the interchangeability policy:
  - Any combination of parts must fit together and work properly, or
  - Some parts may be modified at assembly to fit, or
  - Selective assembly to find a combination of parts which will work together (for example, ultraprecision ball bearings).

How does the policy affect maintenance?

18. Provide correct corner radii, fillet radii, tapers (draft), thickness, thickness variations and junctions or transitions.

19. Material specifications: Have you chosen the most cost-efficient?

20. Study vendor manuals to make your parts conform to vendor processes.

21. Consult your manufacturing or industrial engineers to select processes, adapt your design to them, and to estimate costs.

22. Minimize the variety of raw materials which must be purchased and inventoried.

23. Use commercial components instead of specials. Reference [A8] has a 153-page section on commercial components for use in mechanical design. But for very large production, even a small degree of specialization may be worth it.

24. Pretend that it is your own money.

29.3 LIFETIME COST

1. Minimize the total cost of the product to your customer from the time of purchase until the time of disposal. This cost is the sum of:
- Purchase price
- Operating cost over life (consumables, power, operating labor)
- Maintenance cost over life
- Downtime cost over life
- Minus scrap price at disposal

Industrial and military customers are more rigorous in considering lifetime cost than are most consumers. However, if you watch advertising and read consumer product research magazines, you will discover that many consumers also think about lifetime cost, even if they do not know the phrase.

2. Minimize the net per unit cost of each part, which is the sum of:
- per unit material cost, including scrap
- per unit vendor services cost
- per unit labor cost, with overhead
- average reject and rework cost

3. Minimize the net cost of each assembly, which may not be achieved by minimizing the cost of the individual parts. For example, looser part tolerances may increase assembly labor cost.

4. Minimize the non-recurring cost of the product, which is the sum of:
- R&D and engineering, including market studies
- Tooling
- Setup
- Capital equipment (pro rata, if shared)

5. Consider the cost effect on your company of poor prestige due to poor performance. Hard to measure; a judgment call.

As a designer, you may feel that most of these costs are not your problem. They are. An engineer is paid to design products which will bring profit, and each of these items directly influences profit. Guess what will happen to your own pay, job security, and prestige if your manager knows that you always have all of them in mind.

29.4 WHEN?

There is a continuing flow of new ideas about your design. They come from:

1. Your customer during initial negotiation of your order
2. Your sales department during that negotiation
3. Users and your own sales department as feedback from field experience and from announcements from competitors
4. Other departments, consultants, and advisors in what is lumped as "Concurrent Engineering"
5. Vendors during purchasing negotiations
6. Your managers at any time
7. Your design colleagues and subordinates
8. Mostly from yourself

These ideas come in at any time, often at very bad times. When the author was a very young engineer, he proposed a great new idea for a major improvement right after a set of his drawings went to manufacturing. He was lucky to escape with his job and without physical assault.

The benefit of new ideas, of course, is that they may result in a better product. The malefit[1] of new ideas is cost—in irritation and money and time if you adopt them and in the constant struggle in your own mind to decide if they are really better than what you already have.

Irritation is what happens when you tell a drafter or technician to make a change. The standard responses are, "Do you really need it?" and, "Why didn't you tell me what you really wanted in the first place?" (It took the author years to realize the true answer to the second question: "If I were smart enough, I would have.") Irritation also happens when your manager learns that you are making or proposing a change which will extend his/her schedule and budget.

---

[1] *Malefit* is the author's word for the opposite of *benefit*. He can find no other.

Cost is a trade-off. Are the benefits of the change worth the cost? There are value judgments here, some of which can be calculated but many of which are judgment calls. An entire class of costs is the delays in engineering, manufacturing, sales, and interferences with the schedules of other programs.

On the other hand, it is usually less costly to introduce a change earlier in the program than later.

During design, one can establish deadlines for freezing portions of the design and postpone introducing new ideas until the next design cycle. This can be easier said than done; a really great improvement clamors for use. Judgment.

29.5 PERMITTED CHANGES

Until a product is declared obsolete, there are opportunities to reduce its cost and to improve its performance. But not at any time a good idea occurs to you. The following are some of the permissions and constraints which apply at different times:

1. You may change individual parts, but you may not change the overall design.
2. You may make enough changes to constitute a new model, but not change the overall product.
3. You are designing a new product. Any change is permissible until a design freeze is declared.

A basic approach to scheduling improvements in a released product is to establish phase dates for introducing changes. Changes can then be collected, combined, and revised and then issued in a batch at a scheduled time. There is less confusion, error, and disruption of the factory, and it becomes easier to document what is shipped, since a batch of changes are made starting with a particular serial number.

Advance notice of even proposed changes will enable purchasing and manufacturing to avoid buying or making parts which will become obsolete. Advance notice of changes provides time for tooling.

After drawings are released for manufacture, a continuing stream of detail changes are requested by everyone involved: to correct design errors, to compensate for tooling or manufacturing errors, or to compensate for purchasing problems.

A particularly aggravating problem is a request from manufacturing or purchasing in the form, "This part or material does not quite meet tolerances or specifications, but if we reject and replace it there will be costs in time and money. Can we use it?" If you merely say, "Make it like the drawing," you can cost a lot of money as well as reduce your popularity and your credits for favors in return. If you give an off-hand guess and say OK, you can cost a lot of money if you guess wrong. If you make a thorough study and come up with a correct answer, you can cost some money doing the study but save a lot of money if you can justify acceptance. More judgment.

Another aggravating problem occurs when a fault appears in something already built. The request boils down to, "Fix it but don't change it (FIBDOCI)." You cannot exactly FIBDOCI, but with enough exertion of your ingenuity, you may be surprised at how little "change" will "fix."

29.6 SIMPLICITY

The values of simplicity include cheapness and reliability, ease of understanding, and fewer opportunities for errors and faults. Simplicity itself is not a virtue.

It is not easy to be simple. In art, science, business deal making, diplomacy, mathematics, and design engineering complexity is easy; simplicity is the result of great effort to reduce the complexity.

Your first generation of a design typically has a feature to accomplish each requirement and a structure to hold it all together, the structure itself being made of many parts. Among the many things you do to simplify are:

1. Use fewer parts to perform a requirement.

2. Combine parts to perform more than one requirement in one part or subassembly.

3. Use fewer energy modes (electric, hydraulic, pneumatic, engine, etc.)

4. Use fewer kinds of consumable for example, lubricants.

5. Use fewer manufacturing modes (machining, casting, and so forth)

6. Fold or nest the shapes to be more compact overall.

7. Investigate to see whether some of the requirements are really unnecessary in view of your design so far. Challenge some of the specified requirements to see if they express the real need or benefits of the product. After your initial design work, you may have a deeper insight into what is really beneficial in the product than the person who wrote the specification. You may also be able to offer capability beyond what was originally specified. The specifier had in mind what he/she thought he/she could get and afford. This can do you, personally, a lot of good.

8. Use smooth enclosures which at least give the illusion of simplicity.

9. Make a positive effort to be ingenious. *Wanting* to, and *trying* to, helps to *be* ingenious. (Say "creative" instead of "ingenious" if you prefer the word.)

10. Use commercial devices for effects which would be complicated and expensive to produce yourself. The ball bearing is a common example, but there are innumerable drives, clutches, shock absorbers, structural members and systems, and electro-mechanical and electronic assemblies which can short-circuit much design cost and much product complexity and cost [A8].

The price of this practice is your constant study of other companies' products. An addiction to reading ads in design magazines and marking bingo cards is of tremendous value. However, beware of tying yourself to an unreliable sole source.

29.7 TRADE-OFFS AND TWO-WAY WINNERS

We are accustomed to pay for one benefit by trading off a reduction in another benefit. We sometimes even use the necessity for the reduction as an argument against having to add the first benefit. There is a body of mathematics which optimizes the mix of benefits [A51].

It's not necessarily so.

It is often possible to devise a design which combines two or more benefits. The author coined the phrase "two-way winner" in teaching his young daughter tic-tac-toe. The game is won by finding and marking a square such that there are *two* possible next moves, each a winner.

For example, the ship container. It speeds loading and unloading of ships, protects cargo from damage, prevents pilferage, eliminates warehouses, and provides quick loading and unloading of trucks. A five-way winner which revolutionized transportation.

CHAPTER **30**

Constraints on Design

It would be a wonderful life if we could work in whatever way we felt would produce the best design, but in the real world, we are subject to a large number of constraints. It helps to recognize what they are and learn either how to overcome them or to live within them.

30.1 SPECIFICATIONS AND STANDARDS

Before starting a design, learn the specifications and standards which apply. If you do not, and violate a specification, the cost of redesign may be high. The list is long:

1. U.S. Government specifications:
   - Consumer Products Agency (CPA)
   - Environmental Protection Agency (EPA)
   - Occupational Health and Safety Agency (OSHA)
   - MIL Specs

2. State, county, and municipal specs.

3. Patents.

Patent *claims* specify what you may *not* do without a license from the patent owner. You should order a patent search as part of the specification search before you start design. Theoretically, those patents are a textbook of design ideas, but they make hard reading.

4. Customer's general specs.

Many big companies have general specs which they may or may not "remember" to spell out in their negotiations and purchase order. They have been known to cause severe rework losses to vendors who discover them late.

5. Purchase Order *(P.O.)* "Terms and Conditions."
Typically printed in dark grey small print on a light gray background on the back of the purchase order. More or less balanced by your own "Terms and Conditions" in your quotation and repeated in your P.O. acknowledgement. Good for lawyers' job security.

6. Trade association and professional society specs.
These may be directly or indirectly referred to in the above specs. Many have the force of law, especially if they affect safety.

7. Foreign specs, including, for example:
- Canadian Standards Association (CSA)
- Verband Deutscher Electrotechniker, E.V. (VDE)
- International Electrotechnical Commission (IEC)
- Deutsche Industrie—Norm (DIN)
- International Standards Organization (ISO)

8. Your own organization.
Your own company and department probably have standards for drafting practice, preferred materials and components, manufacturing requirements, procedures, and other things. Some standards apply to your next assignment and some do not. Be sure to find out.

9. Management policies.
Your mangers and your customers have both written and unwritten policies regarding design. If you agree or do not care, they are called policies; if you disagree, you may think of them as prejudices. Either way, you must conform to them as constraints unless you are prepared to wage a campaign for change.

10. There exist standards for materials, components, testing procedures, quality control (especially quality control, or whatever the current slogan calls it; ISO #9000 is the current fashion) and almost everything except the details of your design.

30.2 RESPONSIBILITY

Theoretically your customer and your manager should name all the applicable specs and standards as part of telling you what to design. It is better to do a redundant investigation by yourself than to assume they have done a thorough job. Consider this horror story from the author's experience:

He negotiated a contract for an automatic assembly machine with a major toy manufacturer. The spec had no speed provision, so his company built to its customary speed. Customer engineers arrived for acceptance test, took out their stopwatches, and the machine was junked.

He then learned that their former Chief Engineer had removed a speed spec which had once been part of the overall spec; it was hinted that the man was having mental problems at the time which led to his termination. The customer's president accepted responsibility for the omission and paid for the machine. But in his own mind, the author feels guilty for not going back to the customer in the first place and asking for a speed spec; he should have known it was needed.

30.3 BUDGET AND SCHEDULE

Your design project has a budget and a schedule which you may or may not have influenced. They may constrain the depth of study and experimentation you might like in order to turn out the best-possible design. A loose budget and schedule may permit making a daring experimental design which will be of great value if it succeeds, and if it does not, you can go back to a conservative design. A tight budget and schedule requires conservative design which will certainly work but may be slow, heavy, and expensive.

30.4 CAPABILITIES

Your design may have to be made in a factory with limited capacities in size, accuracy, and so forth. These limitations are constraints on your design.

Your associates and helpers may be fewer and less capable than you would like; they are certainly constraints on your design.

Your model-making and test facilities may be less capable than you would like; they are constraints on your design.

Despite it all, as a retired military officer once told the author,
"You must do the best you can with what you've got."

30.5 SOURCES OF U.S. SPECIFICATIONS AND STANDARDS

See Reference [A8, pp. 239, 240].

CHAPTER 31

People Engineering

31.1 PERSUASION

This is a talk the author gave at the Student Leadership Dinner, Milwaukee School of Engineering, on April 22, 1992. It was well received, so it is offered to you as is, introductory joke and all.

"How to Get a Job Doing Interesting Work and Get Raises and Promotions

Members of the faculty, student leaders:

There was a captain of industry famous for his rousing speeches. They were actually written by a brilliant but very shy speechwriter. One day our speechwriter approached our captain of industry and said,

"Sir, I hate to bother you, but my wife is sick, we have a new baby, prices keep rising, and I haven't had a raise since I came here six years ago. Do you think, perhaps, sir, that maybe you could give me a small increase, please?"

"The trouble with you young people is that all you can think of is money, money, money. I'll give you a raise when you deserve it. Now get out of here and finish my speech for the American Manufacturers' Association."

"Yes, sir. Excuse me, sir."

Black tie dinner. All the other captains of industry. Major generals of the press. I would have been there too, but I was only a second lieutenant of industry.

Our hero opens his Florentine leather notebook and reads:

"I have bad news and good news. The bad news is that your organizations are out of date, your engineering is amateurish, your manufacturing is slovenly, and your managements are incompetent.

"The good news is that I will now tell you how to correct all these faults, so pay close attention."

He turns the page and reads to himself, "OK, you stingy bastard, you're on your own. I quit."

Let us start with some audience participation. Would you please divide into two groups, A and B. There are four qualifications for group A, and you must have all four to qualify:

1. You must be independently wealthy.
2. You must be satisfied to work only on projects you can pay for by yourself.
3. You do not care whether or not anyone uses your ideas or your work.
4. You are not interested in discussing your work with anyone else.

Group B is everyone else.

Members of Group A may now go home; I have a lot to say to Group B.

You members of Group B have the following needs corresponding to the qualifications for Group A:

1. You must persuade an employer to hire you, to promote you, and to pay you increasing amounts of money. Or he/she won't.

2. You must help to persuade customers to pay for your projects by buying your products or by funding your contracts, even if you never meet those customers. If your boss thinks your work is not persuasive he/she will let you go.

3. You must persuade your boss, and your colleagues, and your customers to use your ideas and your work, even after they have paid you to generate those ideas and that work. Or they won't.

4. You must persuade your colleagues to pay attention to your work, to like it, and to help you. Or they will ignore you.

You notice that I use the word *persuade* a lot. Persuasion is what I am going to talk about. Leadership *is* persuasiveness.

Before plunging into the technology of persuasion, let me confirm what you probably already believe: that the most persuasive thing you can do is good work, and lots of it. Most professors, bosses, customers, and colleagues have at least some appreciation of it.

Unfortunately, lots of good work is not enough. People also respond to other motives and influences than the ones they are paid for. If you are to accomplish your four objectives, you must understand these motives and influences and learn how to use them.

There are some among you who instinctively know and practice everything I am about to say. I have met such people, and I cannot tell you how much I envy them. Unfortunately I was one of the majority of engineers who believed that the objective merit of my ideas and my designs was all that counted. Psychological trickery was beneath contempt and certainly beneath me. I have been working to learn better and to act more effectively ever since going from engineering school into industry.

First, let's look at how other people behave and then at how we behave. This will be once over lightly on two of the world's most complex subjects, but a little is better than none at all. If I start some of you thinking along these lines, this talk will have been a success.

I use the word *proposal* to mean an invention, a design, a computation, a plan. Whatever you do in industry, it is a proposal which must be accepted by someone else or it won't happen. Until it is accepted, it is only a proposal. Most academic problems are designed to have right or wrong answers because the objective of these problems is to test whether you have learned something, not what should actually be done. In industry, even a calculation can be contested on the basis of its assumptions or simplifications, especially if the results displease someone.

Other people are motivated by:

1. their self-esteem,
2. by what they judge to be their self interest, and
3. by a variety of doctrines which they believe because they believe.

(How's that for packing a library into a pamphlet?)

If you present a proposed design or plan or invention to Mr. X so that:

1. it makes him feel proud,
2. it makes him judge that it will do him good, and
3. it corresponds to his religion, politics, prejudices, and ethics,

there is a good chance that Mr. X will help along your four objectives.

Conversely, if:

1. you reveal your feeling that he is probably too dumb to appreciate your proposal,
2. show that it will reduce his power and prestige, and
3. that it violates his prejudice that electronics is better than mechanics or vice versa,

he may have limited enthusiasm for your proposal, even though you can absolutely *prove* that it is the very best that can be conceived for the organization, the nation, and the world.

Easy to say. How to do?

1. His self-esteem. Let him know that he is a person who will appreciate the scheme without it having to be spoon fed, that it is partly based on ideas he has expressed, and butter him up any way he will believe. Be ever so respectful. Ask for his opinions and contributions and advice. This is called flattery. You will hate it. I hate it. Do it.

2. His benefits. Show how, directly or indirectly, it will advance *his* four motives. If his judgment is critical, modify your proposal to advance his four motives. A poorer accepted proposal is better than a better rejected proposal. Design for your market.

3. His prejudices, excuse me, his doctrines. If I may be permitted a pedantry, you customer's prejudices are your top specification. The zero'th law of psychology is that people act far more from their feelings than from their logic.

4. How do you know in advance? Listen a lot. Listening is itself flattery, and listening to Mr. X is how you calibrate Mr. X. Watch his body language. Train your sensitivity to other people's feelings. Encourage others to talk about Mr. X. Don't gossip, just listen. Your marketing department calls this market research. Don't sneer at the sales force. If they don't sell; you don't eat.

Now let's look at how *you* behave and how you can make yourself more effective.

There are those who can charm the birds out of the trees. My nephew is one. There are the born leaders who are charismatic and who people obey at once and with enthusiasm. There are the born salespeople. I once met one who made me literally grip the edge of my desk so I wouldn't jump up and shout: "Where do I sign?" even though I *knew* I didn't want his product.

If you are one of these, you may now go out and join Group A; you don't need me and I will attend *your* next lecture.

For the rest of you, I will make only a few of the suggestions I keep trying to teach myself.

1. Real reasons. What are the real reasons you said and did what you said and did? You don't have to be proud of them and you certainly don't have to tell anyone else, but if you learn to understand yourself, you will be able to be far more effective than if you are putting on a show for yourself. Both Socrates and Freud made a big thing of this, one with the expression "Know thyself" and the other with the profession of psychoanalysis.

2. Smile. It can hurt and it can be fake, but others will say yes more.

3. Dress. Part of the local culture of every organization is its unwritten dress code. It is unimportant whether it be no socks and scraggly beard or white shirt and tie. Your appearance is what others see first, and if you send a message of either defiance or effeteness, you may suffer for it.

4. Friendly talk. Don't boast; don't put others down. Affect opinions within the conventional tolerance band and do not hang on the First Amendment. What you really think is your business. People who like you say yes more often than people who do not.

5. Study. There are books and courses on everything from flattery to negotiation.

6. Confine your visible non-conformance to the superior quality and quantity of your work. Really put out your very best. You will feel good, and the more important people will say yes a lot.

As I said; "Once over lightly."

31.2 PERSUASIVE ENGLISH

Until he was 40 years old, the author had the typical engineer's dislike of writing English. *He* wrote drawings and mathematics and provoked his bosses with sketchy and obscure reports. Then he was promoted to the proposal group in Convair. He found himself working with some of the best engineers he had ever known, and they worried about such nonsense as sentence structure and vocabulary! But our products were proposals to NASA and either persuaded them to buy our ideas or we failed. Did he change in a hurry!

You also may want to write information and instructions which your reader can understand easily *because it will benefit you if he / she does.* If so, please consider simple English, as in this book. Use:

1. *Present tense, even for proposals.* ("The beam is strong," not "The proposed beam will be strong.")
2. *First and second person.* ("I read to you," not "The reader reads to the listener." This book is more formal, so the first person is replaced by the third person, "the author.")
3. *Active voice.* ("You hit the ball," not, "The ball is hit by you.")
4. *Imperative mode.* ("Hit the ball," not, "The ball should be hit by you.")
5. *Specific rather than general words.* ("Use the screwdriver," not "Use the tool.")
6. *Short rather than long words.* ("Car," not "Vehicle.")

7. *Just cross out unnecessary words.*

8. *Cut long sentences into short sentences.*

The Bible of clear writing is a very short and very readable paperback, *The Elements of Style* by Strunk and White, published by Macmillan. To go a little deeper, get *The Elements of Grammar* by Shertzer, also a short paperback published by Macmillan.

Is this the Freshman English we all loved so well? It sure is, but now it can do us good!

31.3 POLITICS

Politics is getting benefits, usually competitively. Within an organization the principal benefits sought are:

- Raises
- Promotions
- Budgets
- Perks (perquisites)
- Power (breadth and depth of authority)
- Credit and prestige

Typically, these benefits are assigned by one's higher management, often because the assignment adds the same kinds of benefits to that management.

That recurring word, persuasion, shows up again. A candidate for public office makes persuasive speeches in public and persuasive deals in private. A good engineer persuades by doing lots of superior work.

The author has also seen persuasion by slander, lies, and conspiracies in the most respectable of companies by the most respectable-seeming people.

31.4 EMOTIONAL PROBLEMS IN DESIGN

The great lesson of modern psychology is that we behave primarily for emotional reasons which we argue into pseudorational reasons.

One cannot fully cover human behavior in a list, but here are classes of behavior you will find in others and with which you must try to cope as part of your design effort. (It is up to you to judge whether others find any of these in you, and what to do about it.)

1. *Defensiveness.* Some people feel insecure if they are found to be in error and fight desperately to persuade you that they *are not* in error.

2. *Competitiveness.* If controlled, competitors can be great achievers; if uncontrolled, they can be merely destructive and obstructive.

3. *Conceit.* Whatever is done, it must show how great I am.

4. *Negativeness.* Sometimes a masked hostility. "Your idea is no good because . . . " I rank standard negative responses to new ideas on a scale of one to five under the title of "Innovation Index" in [A3].

5. *Stubbornness.* Devotion to the first idea.

The longer you are an engineer, the longer your own extension of this list. Your success as an engineer depends to a considerable degree on your success in dealing with these behaviors.

31.5 PERVERSITY

There is a tendency to believe that a hostile influence colors our affairs. Thus we say "Just my luck!" when a misfortune occurs, but rarely when good fortune occurs.

In fact, there is great randomness afoot in the world, and more random happenings are adverse than are favorable. In biology, for example, most mutations are harmful and only a few are beneficial, although we owe our genetic success to the survival of the beneficial ones.

In physics, this phenomenon is entropy; the world tends to deteriorate down into randomness, and only human effort can move it up into combinations beneficial to humans.

In daily life, the phenomenon is messiness; consider the continuing effort to keep our desks and our homes reasonably neat.

In engineering, this phenomenon is Murphy's law: "If something can go wrong, it will." In a more general form it is: "Anything that can happen will happen, good and bad." Our entire professional lives are a battle with Murphy. We design organized products out of less-organized materials, and we design these products to resist or overcome random degenerations and failures. We call the resistance "reliability," and we call the overcoming "maintenance."

In applying Murphy's law to design, one can say that anything that can be done can be done badly, but it can also be done well.

31.6 LOGIC

We have been taught that everything fits into categories and that the relationships among categories are the subject of formal logic such as: "If all A's are B's, and if X is an A, then X is a B." Every concept must have a definition which clearly separates it from every other concept, and to use an idea which has no clear definition is an engineering sin. Energy is different from force is different from time.

All of this stems from the 4th century BC Greek philosopher Aristotle, who also taught us that heavy objects fall faster than light objects and a number of dandy ideas about the construction of the heavens, ethics, politics, and much else. At one time, you could get burned at the stake for disagreeing with Aristotle; Giordano Bruno did and Galileo came very close.

In fact, the world is made of spectra more than of categories. A spectrum is a spread of more or less related phenomena. The electro-magnetic spectrum is pretty pure, each element differs from each other only in the wavelength/frequency combination, yet the differences to you and me between gamma rays and broadcast radio waves are quite important.

In human affairs, the spectra of the world have much more blurred edges. How about "good people," "bad design," "strong materials," "reliable," "beautiful," "ethical"? Add some of your own.

The pernicious effects of Aristotelian thinking are not limited to a demand for definitions of ideas we share reasonably well and need no exact definition to use in communication. The worst is trying to reason with spectra as if they were categories. "X is a good person, therefore his/her designs must be reliable."

At last, computers are now programmed to use "fuzzy logic!"

CHAPTER 32

Getting Help

You will get help in your design whether you like it or not, but some help will make you very glad indeed.

32.1 YOUR OWN DEPARTMENT

Within your department, your boss will give you instructions, criticisms, and guidance. The chief drafter and the drawing checker will make changes in your drawings. Administrators will determine your drawing number system. Design assistants, from detailers to colleagues of equal rank, will be assigned to help you and will do so more or less in accordance with your instructions. Model makers and test technicians and engineers will test your designs.

If you are fortunate, there will be other designers with whom you get along well and whom you can consult for ideas and criticisms. Even severely adverse criticism is useful; it gives you an opportunity to recognize and solve problems before they become issues.

(When the author was in the satellite proposal group at Convair/ Astronautics, there was another engineer with the most violent hatred of other people's ideas he had ever met; Frank reacted to a new idea like a mongoose to a snake. However, he was very bright and very well educated, so that if there were a defect in a design, he would instantly and triumphantly nail it. The author always showed Frank his ideas and patiently listened to valid criticisms mixed with sneers. Back to the drawing board; then back to Frank. Sooner or later, Frank would be reduced to pure invective with no rational content at all. Then the author knew he had a winner!)

If you design only a portion of a complex product, you will necessarily confer with the designers of other portions. Airplanes are an extreme case in point.

Other departments—and your customer—have strong and important interests in your design and you should voluntarily consult them at what you think are appropriate times so they do not force redesign later on. It is now common practice for senior managers to insist on design reviews by these departments, not necessarily at the times of your convenience in minimizing the rework they cause. The current buzzword for such involvement is "Concurrent Design." However, please consider this quotation from a letter to the editor of *Machine Design,* September 12, 1991, page 12:

> As a design engineer with one of the largest electronics companies in Japan (and the world), I offer this to the ongoing debate and search for improvement:
>
> Though we have many CAD tools, a smart package cannot compensate for an incomplete understanding of the problem.
>
> Regarding costs, you tend to look down on "bean counters." In Japan, engineering consults with marketing, sales, manufacturing, purchasing, maintenance, and subcontractors. We do not use big names like "concurrent engineering" to describe what we call common sense. We wonder how your readers have been doing their jobs until now. . . .

> (Name kept confidential)
> Mitsubishi Electric
> Inazawa City, Japan

Consider what these other departments are likely to want so you can satisfy them in the first place without having to make a lot of changes later on.

32.2 MANUFACTURING DEPARTMENT

Manufacturing people want to produce your design with a minimum of trouble for themselves. They want to use their existing equipment, techniques, and personnel. They do not want to work to closer tolerances than they now do. They want to be able to inspect partially finished work in order to forestall rework and large-scale rejection. They may be reluctant to subcontract, even to use capabilities they do not have in-house, in order to maintain and augment their empire. (But not always. Some companies have a policy of subcontracting everything they can. You should find out early on.)

32.3 MARKETING DEPARTMENT

Marketing people want to sell the products you design in as large a quantity as possible, at as high a profit as possible, and with the greatest customer satisfaction possible. They want it to work better than the competitors' present and future products, look better, cost less, and require less maintenance. They want features to advertise.

All this may sound facetious, but if you are a designer in a competitive industry (or you would not have a marketing department), you, personally, are in business and should think like a businessperson. Especially if you like job security, raises, and promotions. It is the marketing department which predicts the quantities to be manufactured and sold and the manufacturing rates, which in turn depend on their judgment of your design. Those quantities are fundamental specifications to which you must design.

If you can persuade your marketer to let you meet an actual customer face to face, you will get insights into the real world which are difficult to get secondhand. But the marketer may object to your trespassing on his/her turf.

You may find the personalities in the marketing department more different from your own than those in any other department. They spend their time eating fancy lunches on expense accounts instead of doing honest work, don't they? Of course. But take it from an old entrepreneur, it is the marketing department's selling which makes the company live or die, and with it, your job. The two of you have a lot to teach each other.

32.4 PURCHASING DEPARTMENT

Purchasing people would prefer to buy from their present sources, have a confidential list of sources they do *not* want to buy from, worry about the reliability of new sources, and would prefer to buy materials in the same sizes they now buy because they can get better prices for larger quantities. They worry about buying materials which might suddenly rise in price or become unavailable because of world politics.

32.5 FINANCE DEPARTMENT

These people will get into your act only if your design requires substantial capital investment in new plant, machinery, or tooling. If capital is short or if the product's commercial success is uncertain, they may want you to design for manufacturing with higher unit cost in exchange for lower capital cost.

32.6 R&D DEPARTMENT

You may be required to consult with them in understanding the first-generation designs they produce which you must turn into commercial products. In addition, you may find there invaluable help in solving theoretical problems.

32.7 IN-HOUSE CUSTOMERS

You may design devices for use by your own organization. Such devices include R&D instruments and manufacturing equipment. In this case, you may be able to confer with your in-house customer directly and often.

Customers, both in-house and in the market, may specify design features other than their real needs because they feel that they understand your work and how you should furnish them what they need. It is common for some people to believe that they understand mechanisms because they can *see* them. It is your responsibility to *tactfully* educate them in what is best for both them and you.

32.8 CONSULTANTS

The spectrum of *consultants* ranges from the brilliant expert who instantly solves your desperate problem in a few words, through freelance engineers with a variety of talents and knowledge, to prosaic temporary workers, *job shoppers.* In choosing a consultant, the only substitutes for a knowledgeable referral are your skill at interviewing and plain good luck. A safety net is your ability to terminate the consultant without repercussions from your personnel department. There is more on the subject in [A3, Chapter 10].

In a large organization, you can consult experts and specialists in other groups than your own. Many are in very different professions from yours: pure science, mathematics, medicine, other engineering specialties, and so forth. All you need is some courtesy and a charge number to get startlingly valuable help.

32.9 VENDORS

Some of the most valuable consultants are free: vendor sales engineers. Although they are motivated to sell you their own products, they are usually experts on those products and have seen many applications of them. They are not only free but prompt, and need no paperwork from your organization to bring them in.

Note that it says "sales engineers." Salespeople may or may not be very helpful beyond providing literature, prices, and delivery promises. Manufacturers' representatives, *reps,* are part-time salespeople for many vendors at once and are rarely very helpful in technical matters. As in all categories of people, there are exceptions.

32.10 LIBRARIES

I mention libraries only pro forma. If we include computer-accessible databases, they contain all the knowledge in the world—if you can find the right reference without being swamped.

CHAPTER 33

Design Parameters

It has already been pointed out that EM design engineers deal in a greater variety of materials than most other engineers. It is equally true that EM design engineers deal with a greater variety of parameters than do most other design engineers.

This chapter is a checklist to make sure you have not overlooked a quantity or property which will make your design unsuccessful. It is also a different look at the breadth of EM technology.

33.1 MECHANICAL

- Shape and dimensions
- Stress distribution (stress raisers)
- Strength
- Stiffness/rigidity, and distribution
- Weight, and distribution
- Hardness/wear resistance/abrasion resistance
- Ductility
- Toughness
- Coefficient of friction/lubricity
- Damping
- Creep
- Fatigue

33.2 ELECTRICAL

- Conductivity/resistivity
- Dielectric constant
- Dielectric strength
- Dielectric loss
- Magnetic permeability and saturation
- Eddy current loss
- Hysteresis loss
- Skin effect
- Electro-magnetic forces
- Electro-static voltages
- Corrosion from leakage currents

33.3 THERMAL

- Thermal expansion
- Temperature resistance, high and low
- Specific heat
- Thermal conductivity

33.4 CHEMICAL

- Chemical resistance, including corrosion
- Chemical harmfulness, including pollution by manufacturing process and by product
- Adhesive bondability
- Hygroscopy
- Porosity
- Fading

33.5 BIOLOGICAL

- Toxicity
- Carcinogenesis
- Fungus resistance
- Other microorganism resistance

33.6 OPTICAL

- Transparency, translucence, opacity
- Color
- Refractive index

33.7 ECOLOGICAL

- Effect of discharges on the environment
- Effect of the product on the environment when scrapped [Chapter 27].

CHAPTER 34

Product Classes and Families

34.1 WHY BOTHER?

Fitting your product into classes and families does not design it, so isn't this subject an academic exercise? Not quite.

Classifying your product will help you to design it properly for its production, sale, and use. This classification will also help you to understand the enormous variety of products in the world.

Fitting your design into a family of products made by your company will have many benefits in engineering, manufacturing, marketing, and maintenance.

34.2 CLASSIFICATION BY CUSTOMER

34.2.1 Your Own Organization

Some products are designed for in-house use and can take advantage of that fact. Among these products are:

1. Manufacturing tools, machines, and test equipment
2. Structures, such as shelves
3. Research instruments

In-house products are accessible for modification and maintenance, need not meet some specifications which apply to products to be sold, and usually do not require the cosmetic or artistic treatment necessary for marketed products. For example, a machine which requires experimental changes during development need not be rebuilt before use because of tool and weld marks, rough edges, scratched paint, and the like. Often, documentation required for a product shipped to an outside customer can be dispensed with.

34.2.2 Factories, Commercial Services, and Other Institutions

You can assume that your product will be used under the supervision of technically competent people and maintained by technically competent people, although the actual users may be either unskilled or malicious or both. The duty cycle may be anything from rarely used emergency equipment to 24 hours/day production equipment.

34.2.3 Military

Military customers are a world of their own, with a world of specifications of their own. The requirements usually are based on best quality, long life, arduous service conditions, and maintainability.

34.2.4 Consumers

We are all familiar with the variety of consumer products. Among the things which are less obvious are the government regulations which apply and the exposure to product-liability lawsuits.

34.3 CLASSIFICATION BY USER

The same customer may employ different kinds of actual user. A factory customer may have both unskilled workers and highly skilled technicians.

Different products are treated differently by different users. They should be designed to benefit, benefit from, or be resistant to their users. Users differ in their skill, training, and motivation to *protect the product* and to be *protected from it*. If your product falls into the hands of a user you have not considered, you and your company may be in trouble from loss of prestige, warranty disputes, and product-liability lawsuits, regardless of your fault.

Here is a list of user types you should bear in mind:

1. Unskilled or skilled renter (tools or machinery)
2. Unskilled consumer
3. Unskilled factory worker, perhaps hostile
4. Children, authorized or not
5. Professional who may not be skilled in your product (for example, a physician)
6. Skilled technician, user and/or maintainer

34.4 CLASSIFICATION BY TYPE

34.4.1 Models

1. A *test of principle* (TOP) model may not even resemble the product except in a particular detail. Its purpose is to determine whether you predicted correctly the performance of that detail and, sometimes, to permit measurement of the range within which that performance will be satisfactory. In every other respect, it need not conform to any of the constraints of design.

A TOP model may be used to demonstrate to managers and customers that the principle is a valid one and design should proceed. In electrical work, the term *breadboard* has long been used for TOP models, and its use has spread to mechanical work.

A form of TOP model is the *paper doll* linkages and other structures used by a designer on his board to work out mechanical motions and relationships. Thumbtacks, paper clips, adhesive tape, and staples are all legitimate components of paper doll models.

2. A *visualization model,* or *dummy* or *mock-up,* is used as a three-dimensional layout to help you, your managers, other departments, and your customers visualize a design. Carefully finished mock-ups are used to exhibit the appearance of the design. Accurately sculpted mock-ups are measured for tooling. Some mock-ups are cut away to show internal parts, and some have some moving parts to exhibit their motions.

3. A *test model* is close enough to the real thing to enable a valid test to be run, but may not have the surface finish or other features not relevant to performance. It may have test instruments added which would not be part of the real thing. Most test models are made only once, but some test models are made in small quantities to permit multiple tests.

34.4.2 Single-Article Production

Single-article products, such as a special manufacturing machine or a research instrument, are not models at all since they must conform to all constraints and will be used in the actual services for which they are designed. If the product is to be used in-house, cosmetic blemishes due to experimentation may be forgiven; but if it is to be delivered to a customer, it may be necessary to designate the first article as a test model, scrap it, and reproduce it clean for delivery.

Artistic appearance may be of great importance to a customer. For example, MOBOT Corp. was visited by manufacturing engineers from a famous cosmetics company. They were delighted that we could provide exactly the function they needed and at a quite satisfactory price. They then worried about the aesthetics of our equipment, which was designed for ordinary factories and had no smooth contours and surfaces. All their machinery had been designed with smooth covers and hidden components and was painted pink. PINK? The cosmetics were sold by women who were brought to the factory for inspiration, were met at the airport in a pink Cadillac, and given a tour of the pink factory as part of the visit. Our machine just would not fit in. Even if painted pink. They were not permitted to buy it.

34.4.3 Small-Quantity Production

Techniques of short run-production are described in Chapter 22.

34.4.4 Large-Quantity Production

The challenge in designing a product for quantity production is to design for fast and inexpensive processes and for minimum material cost. In exchange, you are permitted to require a large tooling investment and the processes they permit, such as die casting and plastic molding.

34.4.5 Customized Production

The intent of customized products is to let the customer have a maximum of choice of features with a minimum of extra cost. It is common to price features at very high profit margins. Make the design modular, so he/she can choose which modules he/she wants. Provide mountings for all possible module combinations in a standard base and provide decorative or inconspicuous plugs for unused holes. Provide paint or porcelain enamel finishes so he/she can choose colors and color combinations, with different modules having different colors. The ultimate example, of course, is the automobile and its options.

34.5 PRODUCT FAMILIES

Members of product families use common parts; common or related manufacturing machines, tools, and vendors; similar styling; common marketing and maintenance organizations; common advertising; and similar sales and maintenance literature. In some cases, they can be combined into systems of compatible modules, as with power transmission families.

When you design a new product, take advantage of family relationships. Your marketers are the best source of advice. If your company has a group technology or other standard parts catalog, try to use the parts described in it.

Appendix

Patent Classification

The patent system was established both as a stimulant to invention and as a textbook of technology. For each invention found to be new, the government grants a limited monopoly in exchange for text and drawings teaching the new invention.

There are over four million U.S. patents. To help you find the ones you want, they have been categorized in a system of classes and subclasses in

Manual of Classification of the U.S. Patent Office, which is in two large volumes.

Before entering the manual, consult

Index to the U.S. Patent Classification System.

Both the manual and the index are published by the U.S. Department of Commerce, Patent and Trademark Office, and can be purchased from Superintendent of Documents, U.S. Government Printing Office, Washington, DC 20402.

EMDs are distributed through the U.S. patent system within the following patent classes. This classification scheme was not useful in establishing the classification scheme of this book, since this book is limited to EMDs and the patent classes are not. However if you want to study the subject matter of any portion of this book in greater detail, patents are a wonderful reference and this list is a start.

The following classes were copied from the Index, and each was chosen that including some EMDs. The index lists both classes and subclasses; this search was limited to classes:

CLASS NUMBER	CLASS NAME
181	Acoustics
236	Automatic Temperature and Humidity Regulation
188	Brakes
192	Clutches and Power Stop Control
341	Coded Data Generation or Conversion
342	Communications, Directive Radio Wave Systems and Devices (for example, Radar, Radio Navigation)
340	Communications, Electrical
367	Communications, Electrical: Acoustic Wave Systems and Devices
369	Dynamic Information Storage or Retrieval
360	Dynamic Magnetic Information Storage or Retrieval
381	Electrical Audio Signal Processing Systems and Devices
439	Electrical Connectors
310	Electrical Generator or Motor Structure
338	Electrical Resistors
307	Electrical Transmission or Interconnection Systems
219	Electric Heating
200	Electricity, Circuit Makers and Breakers
174	Electricity, Conductors and Insulators
361	Electricity, Electrical Systems and Devices
337	Electricity, Electrothermally or Thermally Actuated Switches
335	Electricity, Magnetically Operated Switches, Magnets, and Electro-Magnets
324	Electricity, Measuring and Testing
318	Electricity, Motive Power Systems
388	Electricity, Motor Control Systems
323	Electricity, Power Supply, or Regulation Systems
322	Electricity, Single-Generator Systems
191	Electricity, Transmission to Vehicles
313	Electric Lamp and Discharge Devices
314	Electric Lamp and Discharge Devices, Consumable Electrodes
315	Electric Lamp and Consumable Electrodes, Systems

363	Electric Power Conversion Systems
392	Electric Resistance Heating Devices
187	Elevators
373	Industrial Electric Heating Furnaces
228	Metal Fusion Bonding
976	Nuclear Technology
89	Ordnance
355	Photocopying
246	Railway Switches and Signals
346	Recorders
901	Robots
437	Semiconductor Device Manufacturing: Process
116	Signals and Indicators
178	Telegraphy
212	Traversing Hoists
400	Typewriting Machines
251	Valves and Valve Actuation
177	Weighing Scales
242	Winding and Reeling
378	X-Ray or Gamma-Ray Systems or Devices

References and Bibliography

References are grouped as follows. A reference in the text is given by group letter and item number.

[A] Books and articles

[B] Handbooks

[C] Encyclopedias and dictionaries

[D] Magazines with new product announcements

[E] Directories

[F] Catalogs

[G] Trade associations, professional societies, and government agencies

[A] BOOKS AND ARTICLES

[A1] Bell Telephone Laboratories. *Physical Design of Electronic Systems.* 4 vol. New York: Prentice Hall, 1970.

[A2] Dieter, George E. *Engineering Design, A Materials and Processing Approach.* New York: McGraw-Hill, 1983.

[A3] Kamm, Lawrence J. *Real-World Engineering.* New York: IEEE Press, 1991. (Originally published as *Successful Engineering.* New York: McGraw-Hill, 1988). This book is a career guide and has many essays on design and on the life of a professional engineer.

[A4] Ertas, Atila, and Jesse C. Jones. *The Engineering Design Process.* New York: John Wiley, 1993.

[A5] Norman, Donald A. *The Design of Everyday Things*. New York: Doubleday, 1990. (Originally published as *The Psychology of Everyday Things*. New York: Basic Books.)

[A6] Rabinow, Jacob. *Inventing for Fun and Profit*. San Francisco: San Francisco Press, 1990.

[A7] Gordon, J. E. *The New Science of Strong Materials, or Why You Don't Fall Through the Floor*, 2d ed. Princeton: Princeton University Press, 1984.

[A8] Kamm, Lawrence J. *Designing Cost-Efficient Mechanisms*. New York: McGraw-Hill, 1990. Minimum constraint design, commercial components, essays.

[A9] Lee, Thomas H. *Physics and Engineering of High Power Switching Devices*. Cambridge, MA: MIT Press, 1975. A comprehensive but easy-to-read description of circuit breaker technology including arc physics.

[A10] Lilienstein, Fred M. *Magnetic Engineering Fundamentals and Computer-Aided Design Solutions*. New York: Van Nostrand Reinhold, 1993.

[A11] Woodson, Herbert H., and James R. Melcher. *Electromechanical Dynamics*. 3 vol. New York: John Wiley, 1968. A full mathematical treatment.

[A12] Kamm, Lawrence J. "Servo controlled heart and lung machine." *Transactions of the American Society for Artificial Internal Organs*, vol. 5, (1959).

[A13] Calvert, J. Robert. *Mechanics for Electrical and Electronic Engineers*. London: Ellis Horwood, 1992.

[A14] Shigley, Joseph E. *Mechanical Engineering Design*, 5th ed. New York: McGraw-Hill, 1989.

[A15] Logan, Daryl L. *Mechanics of Materials*. New York: HarperCollins, 1991.

[A16] Higdon, Archie, Edward Ohlsen, William Stiles, and John Weese. *Mechanics of Materials*, 4th ed. New York: Wiley, 1985.

[A17] Boothroyd, G. *Assembly Automation and Product Design*. New York: Marcel Dekker, 1991.

[A18] Neubert, Herbert K. P. *Instrument Transducers*, 2d ed. Oxford, UK: Clarendon Press, 1975.

[A19] Young, Warren C. *Roark's Formulas for Stress and Strain*, 6th ed. New York: McGraw-Hill, 1989. This is an all-encompassing handbook of design formulas with their background theory; but you had better have a course in strength of materials behind you.

[A20] Gordon, J. E. *Structures, or Why Things Don't Fall Down*. New York: Da Capo, 1978.

[A21] Swade, Doron D. "Redeeming Charles Babbage's mechanical computer." *Scientific American,* vol. 268, pp. 86–91 (February 1993).

[A22] Nellist, John G. *Understanding Telecommunications and Lightwave Systems.* New York: IEEE Press, 1992.

[A23] Canfield, Eugene B. *Electromechanical Control Systems and Devices.* New York: John Wiley, 1965. Primarily a servo theory book, but it contains descriptions of inertial guidance components together with the mathematics of their operation.

[A24] Gibson, and Tuter. *Control System Components.* New York: McGraw-Hill, 1958.

[A25] Bacon, D. H., and R. C. Stephens. *Mechanical Technology.* New York: Industrial Press, 1990.

[A26] U.S. Air Force, Air Training Command. *Fundamentals of Guided Missiles.* Los Angeles: Aero Publishers, 1960.

[A27] Draper, Charles Stark. "Origins of inertial navigation." *Journal of Guidance and Control,* vol. 4, no. 5, pp. 449–463 (September–October 1981). Theory and history by the principal pioneer in the field.

[A28] Barbour, Neil M., John M. Elwell, and Roy H. Setterlund. "Inertial instruments—where to now?" Draper Laboratory Paper CSDL-P-3182, June 1992. Charles Stark Draper Laboratory, Inc., 555 Technology Square, Cambridge, MA 02139-3563. Presented at the AIAA GN&C Conference, Hilton Head, SC, August 10–12, 1992. A survey and forecast of all forms of inertial instruments.

[A29] Weiner, Norbert. *Cybernetics, or Control and Communications in the Animal and in the Machine.* Cambridge MA: MIT Press, 1961.

[A30] Slater, J. M. *Inertial Guidance Sensors.* New York: Reinhold, 1964.

[A31] Lindsay, J. F., and M. H. Rashid. *Electromechanics and Electrical Machinery.* Englewood Cliffs, NJ: Prentice Hall, 1968.

[A32] Fitzgerald, A. E., Charles Kingsley, Jr., and Stephen D. Umans. *Electric Machinery,* 5th ed. New York: McGraw-Hill, 1990.

[A33] Chironis, Nicholas P. *Mechanisms, Linkages, and Mechanical Controls.* New York: McGraw-Hill, 1965.

[A34] Rubenson, J. G., and G. S. Butterworth. "The inchworm motor." Technical Paper Number 14, American Society of Tool Engineers, 1957.

[A35] *Ney Contact Manual.* J. M. Ney Co., Ney Industrial Park, Bloomfield, CT 06002.

[A36] Machover, Carl. *The C4 Handbook, CAD, CAM, CAE, CIM.* Blue Ridge Summit, PA: TAB Books, 1989. Contains an extensive bibliography.

[A37] Winconek, John E. "Is CAD for you?" *Machine Design,* July 25, 1991, p. 125.

[A38] Petroski, Henry. "Good drawings and bad dreams." *American Scientist,* vol. 79, pp. 104–106 (March–April 1991).

[A39] Petroski, Henry. *To Err Is Human: The Role of Failure in Successful Design.* New York St. Martin's, 1985.

[A40] Glegg, Gordon L. *The Design of Design.* London and New York: Cambridge University Press, 1969. These short books by Glegg were written by a highly experienced design engineer and contain many insights and lessons in design. They are easy and sometimes entertaining to read.

[A41] Glegg, Gordon L. *The Selection of Design.* London and New York: Cambridge University Press, 1972.

[A42] Glegg, Gordon L. *The Science of Design.* London and New York: Cambridge University Press, 1973.

[A43] Glegg, Gordon L. *The Development of Design.* London and New York: Cambridge University Press, 1981.

[A44] Ferguson, Eugene S. *The Mind's Eye: Nonverbal Thought in Technology. Science,* vol. 197, no. 4306, (1977). An excellent study of innovative thinking in engineering, with an extensive bibliography.

[A45] Teicholz, Eric, and Joel N. Orr. *Computer Integrated Manufacturing Handbook.* New York: McGraw-Hill, 1987.

[A46] Nevins, James L., and Daniel E. Whitney, eds. *Concurrent Design of Products and Processes.* New York: McGraw-Hill, 1989.

[A47] Jacobs, Paul, ed. *Rapid Prototyping and Manufacturing.* Dearborn, MI: SME Press, 1992.

[A48] Gore, Al. *Earth in the Balance: Ecology and the Human Spirit.* New York: Houghton Mifflin, 1992.

[A49] Hingoran, N. G., and K. E. Stahlkopf. "High-power electronics." *Scientific American,* (November 1993). Present and predicted future of semiconductor switching in utility systems.

[A50] Nevins, James L., and Daniel E. Whitney, eds. *Concurrent Design of Products and Processes.* New York: McGraw-Hill, 1989.

[A51] Arora, J. S. *Introduction to Optimum Design.* New York: McGraw-Hill, 1989.

[A52] Holzman, Gerard J., and Bjorn Pehrson. "The first data networks." *Scientific American,* (January 1994).

[A53] Bryzek, Janusz, Kurt Petersen, and Wendell McCulley. "Micromachines on the march." *IEEE Spectrum,* (May 1994).

[A54] Comerford, Richard. "Mecha . . . what?" *IEEE Spectrum,* (August 1991).

[A55] Cheney, Margaret. *Tesla, Man Out of Time.* Englewood Cliffs, NJ: Prentice Hall, 1981.

[B] HANDBOOKS

[B1] Avallone, E. A., and T. Baumeister III, eds. *Marks Standard Handbook for Mechanical Engineers,* 9th ed. New York: McGraw-Hill, 1987.

[B2] Walsh, Ronald A. *Electromechanical Design Handbook.* Blue Ridge Summit, PA: TAB Books, 1990. An entry-level book on mechanical design applicable to electro-mechanical devices.

[B3] Parmley, Robert O., ed. *Mechanical Components Handbook.* New York: McGraw-Hill, 1958.

[B4] Shigley, Joseph E., and Charles R. Mischke. *Standard Handbook of Machine Design.* New York: McGraw-Hill, 1988. This is a comprehensive handbook containing encyclopedia information, design formulas, and data.

[B5] Bralla, James G. *Handbook of Product Design for Manufacturing.* New York: McGraw-Hill, 1986.

[B6] Fink, Donald G., and H. Wayne Beaty, eds. *Standard Handbook for Electrical Engineers,* 12th ed. New York: McGraw-Hill, 1993.

[B7] Norton, Harry N. *Handbook of Transducers.* Englewood Cliffs, NJ: Prentice Hall, 1989.

[B8] Society of the Plastics Industry, Inc. *Plastics Engineering Handbook.* New York: Reinhold, 1976.

[B9] *Handbook of Chemistry and Physics.* Boca Raton, FL: CRC Press, published annually. Mostly data of pure science, but some of it is useful in EM design.

[B10] Brady, George S., and Henry R. Clauser. *Materials Handbook,* 12th ed. New York: McGraw-Hill, 1986. This handbook describes many kinds of material in addition to the usual engineering metals and plastics.

[B11] Powell, Russell H., ed. *Handbook and Tables in Science and Technology,* 2d ed. Phoenix, AZ: Oryx Press, 1983.

[B12] Perry, R. H., and D. W. Green. *Perry's Chemical Engineers' Handbook.* New York: McGraw-Hill, 1984.

[B13] Bolz, Ray E., and George L. Tuve, eds. *CRC Handbook of Tables for Applied Engineering Science.* Boca Raton, FL: CRC Press.

[B14] Craig, B. D., ed. *Handbook of Corrosion Data.* Novelty, OH: American Society for Metals, 1989.

[B15] McPartland, . *McGraw-Hill's National Electric Code Handbook.* New York: McGraw-Hill. Also see [G3].

[B16] *Materials Selector 1994 (Machine Design* magazine annual issue). Cleveland, OH: Penton Publishing. (Formerly designated an annual issue of *Materials Engineering* magazine.) A detailed and well-organized data handbook on engineering materials. It also contains comparison tables of great value in choosing materials. I think it is one of the most valuable handbooks a designer can have. Also see [D3].

[B17] Parr, E. A. *Industrial Control Handbook.* vol. 1, *Transducers.* New York: Industrial Press, 1987.

[B18] Bolz, Roger. *Production Processes: The Productivity Handbook.* New York: Industrial Press, 1981.

[B19] Oberg, E., F. D. Jones, and H. L. Horton. *Machinery's Handbook,* 23d ed., New York: Industrial Press, 1988. This is an extremely comprehensive handbook of data on mechanical design.

[B20] Cheremisinoff, Nicholas P. *Unit Conversions and Formulas Manual.* Ann Arbor, MI: Ann Arbor Science Publishers, 1980.

[B21] Woodson, Wesley E. *Human Factors Design Handbook.* New York: McGraw-Hill, 1981. This 1000-page book contains encyclopedic text and design data.

[B22] Croft, Terrel, and Wilford I. Summers, eds. *American Electrician's Handbook,* 12th ed. New York: McGraw-Hill, 1992. Elementary descriptions of many electro-mechanical devices, particularly division 4.

[B23] Smeaton, Robert W., ed. *Switchgear and Control Handbook.* New York: McGraw-Hill, 1977.

[B24] Chironis, Nicholas P., ed. *Mechanisms and Mechanical Devices Sourcebook.* New York: McGraw-Hill, 1991.

[C] ENCYCLOPEDIAS AND DICTIONARIES

[C1] *McGraw-Hill Encyclopedia of Science and Technology,* 6th ed. New York: McGraw-Hill, 1987.

[C2] Meyers, Robert A., ed. *Encyclopedia of Physical Science and Technology,* 15 vols. New York: Academic Press, 1987.

[C3] *Encyclopedia Britannica.*

[C4] *Encyclopedia Americana.*

Both [C2] and [C3] have comprehensive articles in engineering and science.

[C5] *Random House Dictionary of the English Language* (unabridged).

[C6] *Webster's Third New International Dictionary* (unabridged). An unabridged dictionary serves as a miniencyclopedia of engineering and science as well as of language in general.

[C7] Lapedes, D. N., ed. *McGraw-Hill Dictionary of Scientific and Technical Terms,* 2d ed. New York: McGraw-Hill, 1978.

[D] MAGAZINES WITH NEW PRODUCT ANNOUNCEMENTS

[D1] *Materials Engineering,* Reinhold Publishing Co., 600 Summer St., Stamford, CT 06904, 203-348-7531. Free if you qualify.

[D2] *Design News,* Cahners Publishing Co., 275 Washington St., Newton, MA 02158, 617-964-3030. Free if you qualify.

[D3] *Machine Design,* Penton Publishing, 1100 Superior Ave., Cleveland, OH 44114, 216-696-7000. Includes annual *Materials Selector Issue* [B16]. Free if you qualify.

[D4] *New Equipment Digest,* Penton Publishing Co., 1100 Superior Ave., Cleveland, OH 44114, 216-696-7000. Free if you qualify.

[E] DIRECTORIES

[E1] *Thomas Register of American Manufacturers,* Thomas Publishing Co., 1 Penn Plaza, New York, NY 10119, 212-695-0500.

[E2] *Guide to American Directories,* B. Klein Publications, Coral Springs, FL.

[E3] *Sweet's Mechanical Engineering Catalog,* Sweet's Catalogs, McGraw-Hill, 1221 Avenue of the Americas, New York, NY 10020, 212-512-4442.

[E4] *MacRae's Blue Book,* MacRae's, 817 Broadway, New York, NY 10003, 212-673-4700.

[E5] *The Standard Periodical Directory,* Oxbridge Communications Inc., 150 5th Ave., New York, NY, 10011, 212-741-0231.

[F] CATALOGS

[F1] McMaster-Carr, P.O. Box 54960, Los Angeles, CA 90054, 213-945-2811.

[F2] Edmund Scientific Co., 4104 Edscorp Bldg., Barrington, NJ 08007, 609-573-6266.

[F3] Newark Electronics, 4801 N. Ravenswood Ave., Chicago, IL 60640-4496, 312-784-5100.

[F4] Small Parts Co., P.O. Box 381966, Miami, FL 33238, 305-751-0856.

[F5] PIC Design Co., P.O. Box 1004, Benson Rd., Middlebury, CT 06762, 203-758-8272.

[F6] Winfred M. Berg Inc., 499 Ocean Ave, East Rockaway, NY 11518, 516-599-5010.

[F7] Stock Drive Products, 2101 Jericho Turnpike, New Hyde Park, NY 11040, 516-328-3300.

[F8] Value Plastics Inc., 3350 East Brook Dr., Fort Collins, CO 80525, 303-223-8306.

[F9] Boston Gear, 14 Hayward St., Quincy, MA 02171, 800-343-3353.

[F10] L.A. Rubber Co., P.O. Box 23910, Los Angeles, CA 90023, 213-263-4131.

[F11] Linear Industries Inc., 1850 Enterprise Way, Monrovia, CA 91016, 818-303-1130.

[F12] W. W. Grainger Inc., 2738 Fulton St., Chicago, IL 60612, 312-638-0536.

[F13] Minarik Electric Co., Suite 101, 165 E. Commerce Dr., Schaumburg, IL 60173, 312-885-9337.

[F14] *Electronic Engineers' Master Catalog,* Hearst Business Communications Inc., 645 Stewart Ave., Garden City, NY 11530, 516-227-1300.

[G] TRADE ASSOCIATIONS, PROFESSIONAL SOCIETIES, AND GOVERNMENT AGENCIES

These organizations publish books, specifications, standards, and periodicals relevant to electro-mechanical devices and systems. Certain publications are named and underlined.

[G1] IEEE Components, Hybrids, and Manufacturing Technology Society, *Transactions.*

[G2] U.S. Patent Office, Superintendant of Documents, U.S. Government Printing Office, Washington, DC 20402. *General Information Concerning Patents; Manual of Classification.*

[G3] National Fire Protection Association, Batterymarch Park, Quincy, MA 02269. *National Electric Code.* The National Electric Code is a book of safety rules for the design and installation of electrical equipment. Also see [B15].

[G4] National Electrical Manufacturers' Association (NEMA), 2101 L St. NW, Suite 300, Washington, DC 20037.

[G5] Underwriter's Laboratories (UL), 333T Pfingsten Rd., Northbrook, IL 60062.

[G6] Institute of Electrical and Electronics Engineers (IEEE), 345 East 47 St., New York, NY 10017.

[G7] American Society of Mechanical Engineers (ASME), 345 East 47 St., New York, NY 10017.

[G8] Instrument Society of America, Box 12277, Research Triangle Park, NC 27709.

[G9] American Society for Testing And Materials (ASTM), 1916 Race St., Philadelphia, PA 19103.

[G10]Society of Automotive Engineers (SAE), 400 Commonwealth Ave., Warrendale, PA 15096.

[G11] Federal and Military Specifications, U.S. Naval Supply Depot, 5801 Tabor Ave., Philadelphia, PA 19120. *MIL Specifications.*

[G12] Edison Electric Institute, 750 Third Ave., New York, NY 10017.

[G13] Electric Power Research Institute (EPRI), Menlo Park, CA.

[G14] American National Standards Institute, Inc., 1430 Broadway, New York, NY 10018. *Catalog of American National Standards.*

[G15] American Society for Metals (ASM), Metals Park, OH 44073.

Index

TJ 163.12 .K35 1996
Kamm, Lawrence J.
Understanding electro-
mechanical engineering